The Restless Sun

The Restless Sun

Donat G. Wentzel

Smithsonian Institution Press
Washington, D.C.
London

Editor: Vicky Macintyre
Coordinating Editor: Leigh Alvarado

LIBRARY OF CONGRESS CATALOGING-IN-PUBLICATION DATA

Wentzel, Donat G., 1934–
   The restless sun.

   (Smithsonian library of the solar system)
   Bibliography: p.
   Includes index.
   1. Sun. 2. Astrophysics. I. Title. II. Series.
QB521.W46 1989   523.7   88-18491
ISBN 0-87474-982-4

British Library Cataloging-in-Publication Data Available

Manufactured in the United States of America
10 9 8 7 6 5 4 3 2 1
98 97 96 95 94 93 92 91 90 89

∞ The paper used in this publication meets the minimum requirements
of the American National Standard for Permanence of Paper for
Printed Library Materials Z39.48-1984

To my mentors,
my grade school teacher, Alfred Surber,
my father, Gregor Wentzel, and
S. Chandrasekhar.

Contents

Preface

A New Sun is the title of a book by John Eddy published in 1979. It presents pictures of the Sun produced during the Skylab manned space missions in 1973–74. Many were taken with X-ray cameras mounted on the outer skin of Skylab (plate 1). One spectacular X-ray picture appears as plate 2. Such pictures surprised even the experts. They caused such drastic changes in our understanding of the Sun that *A New Sun* was a particularly appropriate title.

Pictures are an essential ingredient of solar studies. Many of the new ideas of the past two decades are best explained by pictures. The technical terms that may be necessary to express these ideas are most simply introduced through pictures. The pictures in this book portray our knowledge of the Sun. Just as the mention of John F. Kennedy evokes a mental panoply of historical photographs, so the mention of solar coronal loops or magnetic arches later in this book should evoke Skylab pictures.

Solar science abounds with discoveries of new phenomena because it has been blessed with many advances in high technology. For example, photographs have been improved because the mirror in one of the solar telescopes has been electronically corrected to remove the fuzziness caused by the Earth's atmosphere. Radio observations have benefited because new electronics make their radio receivers ever more sensitive, so sensitive in fact that many radio observations can be converted into photographs, which are more easily appreciated than the numbers and graphs usually obtained with a radio telescope. A host of space experiments have acquired information on solar radiations that are never observable from the ground. Skylab X-ray data, in particular, have been converted into spectacular photographs.

The photographs in this book represent an enormous range in observational capabilities, many of which were nonexistent merely 20 years ago. Since then, Skylab has brought about a revolution in our thinking, which still plays a central role in solar science even though it happened well over a decade ago. The Solar Maximum Mission launched in 1980 has brought further discoveries. Recent advances have also been dramatic. Indeed, this book portrays solar science as a very active field of study, as active as the Sun itself.

Accuracy is the hallmark of celestial mechanics: Measurements from Earth can pinpoint the position of a space probe beyond Jupiter to within 1 kilometer. Accuracy is not yet a characteristic of solar studies. Pictures tend to be qualitative. They portray the essential phenomena. They portray the Sun as forever restless. Dynamic changes on the Sun are so diverse, their temperatures and their durations vary so widely, that a rough value of a quantity is usually of greater interest than its precision. An "order of magnitude" evaluation is usually sufficient: Is an object a hundred, thousand, or a million kilometers in size? Does an explosion reach temperatures of merely a million degrees or might it approach a billion degrees? Whether an explosion occurs at 10,000, a million, or 100 million degrees is essential information because completely different observing methods are appropriate for such different temperatures. Certainly one tries to evaluate the temperatures more precisely, but distinctions between, say, 5,000 and 6,000 degrees are usually relegated to secondary, technical considerations. In this book, temperatures are given in degrees Kelvin, but, since all temperatures of interest are in the thousands and the millions of degrees, the difference of 273 degrees from the Centigrade (Celsius) scale is immaterial. Similarly, distances are stated in kilometers, but the difference between a mile and a kilometer is not of vital concern.

Many people think that scientists spend all their time poring over their books and data. Indeed not! They spend much of their time planning how to obtain new data, often through the most recent advances in observing technology, in the hope of answering newly asked scientific questions. But new technology, particularly space technology, is expensive to use. New observations must be carefully planned in order to obtain the maximum science from the available budgets. Solar astronomers, like most astronomers and perhaps more than other scientists, frequently have to participate in the long-term planning of their observations. Computer-corrected mirrors for solar telescopes, specialized receivers for radio telescopes, and especially space experiments take years to plan and build.

Important instruments on the ground may cost a million dollars; space experiments may cost hundreds of millions of dollars. To obtain funds of this magnitude, one must convince colleagues, panels of experts, committees of scientists from other fields, relevant officials at the funding agencies, and ultimately Congress and the public that such expenditures are worthwhile. Many scientific, practical, and political questions must be answered. Are the proposed observations significant for the discipline of solar studies? Will the interpretations be reliable? Are they important for other aspects of astronomy, for our understanding of the Earth, or for physics? Will they help us deploy a space station safely in the vicinity of the Earth? Will the results expand our view of the Universe? Can the experiments be built within the budget and within the allotted time? Will the experiments produce popularly attractive pictures that help establish good public relations? These questions pervade solar science continually. They appear throughout the book and provide a focus for the last chapter.

I thank my wife, Maria, and also Jay Pasachoff and an anonymous reviewer for many useful comments. This book was written independently of my then employer, the National Science Foundation.

Introduction:
The Many Time Scales of the Sun

Time Scales

Nothing in solar studies covers more orders of magnitude than the time scales on which the Sun changes. Some solar phenomena may last only a tenth of a second, others billions of years. A good way to keep track of them is to arrange them according to the times involved. The phenomena listed in table 1, and described in this introduction, have attracted the most public attention. All are discussed in later chapters, under the four main topics around which the book has been organized: the layered Sun, structure and evolution; solar activity; flares; and the relation of these topics to the Earth and other stars.

Flares

Solar flares are explosions. Their energies dwarf those of any explosions on Earth. For perhaps a tenth of a second, explosion temperatures may reach several hundred million degrees! "Moderate" temperatures like 20 million degrees may last for a minute or more. At these temperatures, most flare radiation is in the form of X rays, which cannot pass through our atmosphere. That is why flares are best observed from space. Plate 2 shows an X-ray picture of the Sun taken by Skylab during a flare. The flare appeared as an intense tiny source of X rays near the middle of the solar disk.

The flare explosion expands and "cools" rapidly. It then appears as a bright region on the Sun's disk that can be photo-

Table 1. Duration of Solar Events

0.1 seconds	Flares at 100 million degrees, radio bursts
1 minute	Flare "explosive" phase
10 minutes	Motions on solar surface
Hours	Coronal "transients" to interplanetary space
Days	Maturation of sunspots, spot groups
Weeks	Solar rotation, "combing" of corona
11, 22 years	Solar cycles
Centuries, millennia	Maunder minimum, climate changes, radius changes (?)
20 million years	Solar turn-on, major ice ages (?)
600 million years	Evidence for ancient solar cycle
5 billion years	Age of Sun, Earth; time until solar senility

graphed from the ground. The bright region expands explosively for one or a few minutes. At maximum, a large flare such as the one in figure 1 may cover about 1 percent of the solar disk, attaining a length of about 12 Earth diameters! Flares looking much like figure 1 also occurred in 1959 and 1966.

The most intense flares can actually be observed with the unaided (but protected) eye. In fact, flares were first discovered when Richard C. Carrington observed such a flare on September 1, 1859. He was measuring the position of an unusually large spot on the solar disk when he saw a pair of brilliant white ribbons flash across the spot. The flare faded from view in a mere five minutes.

Flares also cause much radio emission. If a radio telescope had converted the flare's radio signals into sound, the sound would have resembled that in a shooting gallery. The flare "gallery" is 150 million kilometers away from us!

The enormous temperature of a flare and its repeated shotlike behavior resemble what is observed in fusion machines, the machines used for research toward the peaceful use of nuclear fusion energy. Our observation of solar flares may indicate whether we

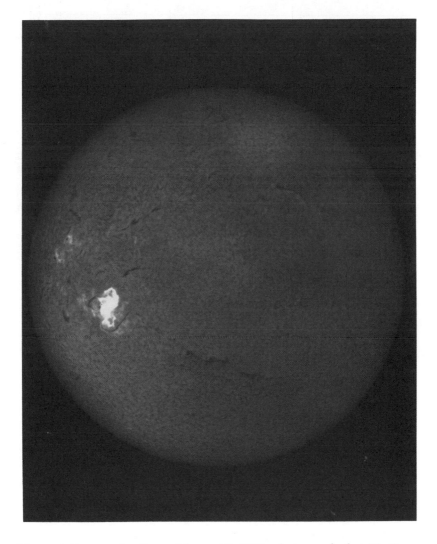

Figure 1. Large solar flare of August 7, 1972, photographed at Big Bear Solar Observatory, California. The flare appears as an intense brightening, with a length of about 150,000 kilometers, or 12 Earth diameters, covering almost 1 percent of the disk. Typically, such a flare grows explosively during a few minutes and takes about an hour to fade away. The shape of a "sea horse" has appeared in several major flares. A close-up of this flare appears in Figure 11.1. (Big Bear Solar Observatory)

really understand what will happen in planned fusion machines on Earth.

Spots

Sunspots look like dark blemishes on the Sun. Those visible to the naked eye have been observed and recorded for many centuries. However, such observations have been restricted to the rare occasions when sunspots are unusually large and when the Sun is sufficiently obscured by haze or dust to be viewed without damaging the human eye. In naked-eye sightings of the past, no pattern was perceived and the sunspots were explained as terrestrial or planetary phenomena. There was no reason to assign the spots to the Sun itself. Up to the time of Galileo, there was also no desire to do so, given the then-traditional view of a perfect, unblemished Sun.

Galileo's observations (figure 2) forced a change of opinion. Galileo observed the spots regularly, using his telescope to project an image of the Sun. He found that the spots appeared to advance across the solar disk, day by day. Since he could observe sunspots with ease, he made a daily map of their positions on the face of the Sun. These maps convinced him that the spots were really a part of the Sun, but that they appeared to move across the face of the Sun because the Sun rotates. The Sun rotates once in about four weeks. The notion that the spots are actually solar blemishes was contrary to established dogma of the time. The recognition of the imperfectness of the Sun contributed to the rise of the heliocentric theory of the solar system.

Sunspots change slowly in the course of several days or weeks. Single spots may appear tiny and then grow and split. They usually appear in groups. Figure 3 is a modern photograph of an unusually large group of sunspots. Day-to-day photographs would show many changes in the small spots. When spot groups first become visible, they tend to contain only a few simple spots, but with time these grow, become more complex, and split into more spots, which then dissolve and become unobservable in the course of several weeks. Their behavior is quite unpredictable, especially during the two weeks of invisibility.

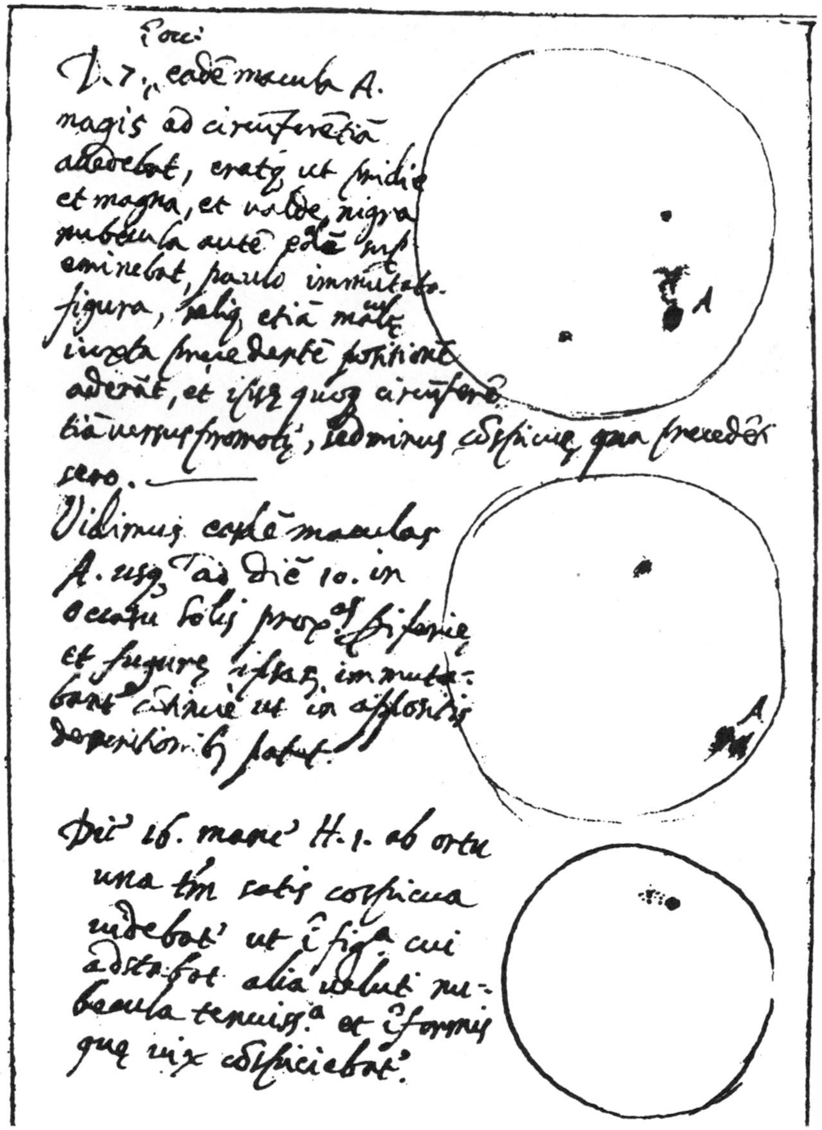

Figure 2. Galileo's daily record of sunspot positions. The apparent regular progression of the spots across the solar disk convinced Galileo that the spots are a part of the solar surface and rotate with the Sun. (Yerkes Observatory)

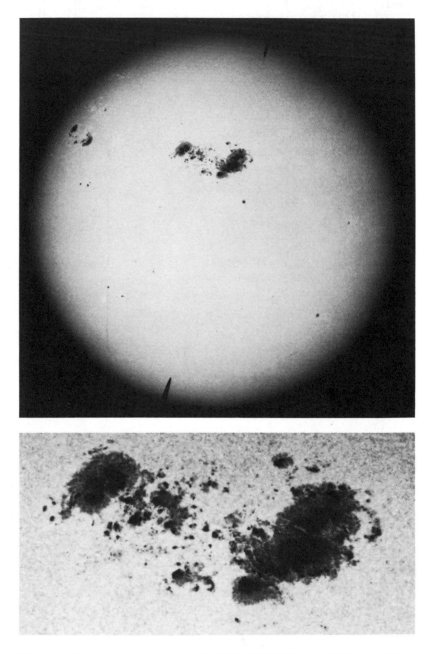

Figure 3. Large sunspot group, April 7, 1947. (Mount Wilson and Las Campanas Observatories, Carnegie Institution of Washington)

Some groups reappear for several rotations; others disappear completely while out of view.

Solar Cycles

The sunspot cycle was discovered in 1843 by Heinrich Schwabe. He recognized a cycle in the number of sunspots visible at any one time on the Sun. In some years many spots are present. Then follow years with few spots, followed again by years with many. Years of maximum sunspot numbers, more simply termed years of solar maximum, are separated by about 11 years. Thus one talks of an 11-year sunspot cycle. By 1843, spots had already been observed routinely for two centuries, yet the sunspot cycle was not recognized during all that time! Today the sunspot cycle is not only well known but is also popularly associated with cycles in climate (a subject of hot debate) and even cycles in sociological phenomena, such as the length of women's skirts.

Sunspots are magnetized and are arranged according to their magnetic polarities. The magnetic pattern takes about 22 years to complete a cycle, twice the ordinary sunspot cycle. Some climate patterns show a 22-year cycle. Perhaps one causes the other, but if so, an extraordinarily long chain of events is involved that is still being unraveled.

Sunspots are relatively dark. Do we receive less sunlight when there are large spots? Recent measurements with the Solar Maximum Mission spacecraft tell us that a large spot group reduces sunlight by about 0.1 percent. No direct terrestrial consequence of this deficiency has yet been determined. Nevertheless, large sunspots are watched carefully because flares tend to occur near large spot groups, and large flares have clear terrestrial consequences. Large flares occur preferentially when spot groups grow rapidly and become more complex. Since new spot groups appear and grow most frequently near the time of sunspot maximum, large flares occur most frequently near the time of solar maximum.

Fortunately for solar physics, the terrestial consequences of solar flares translate to substantial financial support for flare research. A conspicuous example of flare research is the Solar

Maximum Mission, launched on February 14, 1980, in association with the solar maximum of 1980–81. The Solar Max or SMM successfully observed flares beyond all expectations. However, it became famous primarily because it was the first satellite designed to be repaired in space, and it actually needed such repair.

Corona

Every year or two, the Moon covers the Sun entirely, at least as seen from selected locations on Earth. During those few minutes of solar eclipse, an eerie glow surrounds the covered-up Sun (plate 3). The glow is called the corona, from the Latin for "crown." The phenomenon has been known for millennia. But no interpretation was possible until atomic physics was developed sufficiently, about five decades ago. Surprise: The corona is over a million degrees hot! One theory for the heating of the corona survived for twenty years, far longer than most theories in astronomy. But it fell by the wayside with the arrival of the space age.

Space experiments can measure solar X rays. Suitable cameras can produce X-ray images of the Sun that appear much like ordinary photographs. X rays are a much better measure of million-degree gases than is the visible light recorded in eclipse photographs. Moreover, the X rays come only from the hot corona, not from the solar surface. X-ray cameras are not blinded by the solar surface as are ordinary cameras. Thus X-ray images of the corona can be taken at any time and are not restricted to solar eclipses.

Dramatically new observations were obtained from Skylab. This first U.S. space station was occupied by three teams of astronauts for about three months each during 1973–74. The astronauts had many duties, but solar observations and specifically X-ray imaging were a significant component of their work. Their frequent and long-term observations improved our understanding of the Sun to such an extent that even the experts were astounded.

One surprise was the temperature of flares, which often reach well beyond 100 million degrees.

A second surprise: The solar corona turned out to be much more structured than was expected from the views available during eclipses. Plate 4 shows one of the more dramatic X-ray pictures provided by Skylab. There are seven centers of emission, connected by "loops." The loops are the unexpected feature. They constitute giant electrical circuits, up to 300,000 kilometers long!

Yet another surprise: The solar corona turned out to be much more dynamic than expected from the occasional "snapshots" at times of eclipse. Skylab could detect the ordinary corona much further from the Sun than is possible from the ground during an eclipse. The Skylab photographs revealed enormous bubbles of coronal gas that seem to lift off from the Sun for no obvious reason and during the next few hours travel outward into interplanetary space. They have received the descriptive name "transients" (plate 5).

When the images from the Solar Maximum Mission are combined into a motion picture, transients appear to comb out the corona. The transients remove most of the coronal gas from the Sun in merely some three weeks. At the same time, the gas is somehow replenished from below. This is far different from the static corona envisaged earlier.

Sunspots, flares, and transients are part of "solar activity," a term used to indicate that the seemingly quiet Sun involves continual motions. In fact, on film the solar surface looks rather like a mass of writhing and wiggling worms. The patterns in the seething gases of the solar surface appear and disappear every few minutes. The Sun is forever restless. However, the "quiet Sun" is an abstraction that also helps humans to understand the Sun and is therefore retained in the first part this book (chapters 1–4).

Is the Sun unusual in having a hot corona and much activity in the form of spots and flares? Is it unusual in emitting X rays? Not at all. Most other sunlike stars also emit X rays. Some of the stars emit X rays much more copiously, which indicates very active stellar coronae. Other stars emit X rays so weakly that we can barely detect them with present equipment. If the Sun were at typical stellar distances, it would be in this weaker class of stars.

Effects on Climate?

The solar cycle is somewhat irregular. The intervals between solar maxima vary between about 9 and 13 years. "Eleven-year cycle" has become a common name for this phenomenon. The number of spots also changes from cycle to cycle. The number of spots during some maxima may be twice the number at other maxima. Indeed, it appears that sunspots may disappear almost entirely, for several decades. An example was the "Maunder minimum," a period from A.D. 1645 to 1715 during which almost no sunspots were recorded. This was also a period in which the climate in North America and Europe was unusually cold. It has been likened to a mini ice age. Might the two phenomena be related? At first, skeptics doubted that the solar cycles would change this drastically. Perhaps the lack of recorded sunspots was merely related to the methods used to document them. John Eddy checked historical records for numerous phenomena associated with sunspots. He showed that these also were nearly absent. Seven solar cycles were skipped entirely. Whether that caused the mini ice age is still open to debate.

If solar cycles may occasionally languish, they can also be hyperactive. The evidence comes from naked-eye recordings of sunspots throughout history. These observations reflect sunspot groups at least the size of the largest recorded in the present century. They also indicate that in some centuries solar maxima were unusually active. These extremes in the solar cycle have attracted much interest, because they may be related to changes in the Earth's climate and thus may have influenced the development of human civilization. Remarkably, a solar cycle of about the present length, 9 to 12 years, appears to have operated even 600 million years ago. Ancient lake deposits in Australia indicate such a cycle. When these deposits were laid down, the solar cycle apparently was extreme enough to influence the local climate, at least enough to cyclically change the water runoff that caused the deposits.

Some precise measurements of the solar radius suggest that the Sun's radius has changed by small fractions in the course of recent decades and the past few centuries. The measurements are quite controversial, and other measurement disagree, but they attract much attention because a decade is short enough to influence government planning. Small changes in radius may imply similar changes in solar energy reaching the Earth. Just how such changes may alter the complex balance of the terrestrial environment is still unknown. But climatic changes caused by the Sun can only grow in importance as mankind expands into the ever more sensitive and hostile unpopulated environments left on the Earth.

It is possible that the Sun's brightness changes not just over weeks, decades, and centuries, but also over millions of years. Solar energy is produced near the center of the Sun and takes millions of years to reach the surface. It takes 20 to 100 million years for a change in the deep solar interior to emerge and become observable at the surface. We know that major ice ages occur at intervals of about 100 million years. Much debate centers on the question, Are major ice ages related to million-year changes in the solar interior?

Solar Lifetime

On the longest time scale, the Sun is using up the nuclear fuel near its center, which produces the energy radiated into space. The nuclear fuel, hydrogen, will be exhausted when the Sun is about 10 billion years old. One regards 10 billion years as the lifetime of the Sun. The Sun is now about halfway through its life. Astronomers observe many stars like the Sun that are currently in various stages of birth, life, or death.

Early solar life lasted a relatively short period—about 20 million years—during which the star as a whole settled into a steady state. The Earth was also formed at that time, about 4.6 billion years ago. There followed a period of about a billion years in which solar spot and flare activity was intense. Judging from other stars, observed in their youth, present solar activity is a mere vestige of the Sun's activity in its youth. Fortunately for life on Earth, the Sun has been and will remain rather dull for a long time. But change is inevitable. In about 5 billion years, the death of the Sun will eliminate all life from Earth. First the

Sun will expand to scorch the Earth; then it will shrink so that the Earth becomes thoroughly frozen.

The solar center holds a mystery. An experiment was built to detect the particles—called neutrinos—that are emitted from the solar center. Surprisingly, only about one-third as many neutrinos were observed as predicted. This discrepancy may have something to do with the fact that we do not fully understand some of the basic aspects of physics. The same physics that might explain the results of the neutrino experiment also may provide many answers to questions about the origin of the universe. New experiments on solar neutrinos now being designed may, in effect, test current cosmological theories.

The Sun is neither immutable nor perfect in the classical sense. A study of the Sun provides many scientific surprises and yields new understanding of the Earth and of the Universe. It should induce some modesty in our belief that we humans can comprehend stars, galaxies, quasars, and the Universe.

PART I

THE LAYERED SUN: STRUCTURE AND EVOLUTION

Chapter 1

Terrestrial Prejudices:
The Solar Fires

The Sun provides most of the energy by which plants and animals survive on Earth. Mythology speaks of the solar fire. Is the Sun really burning? Might the Sun consist of wood? Coal? Oil? Today we know the answer is no. Nevertheless, the concept of a solar fire seemed quite reasonable right up to the nineteenth century.

What can we learn about the Sun without traveling toward it, without sampling it, without modern knowledge of atoms or nuclear energy?

Seventeenth Century: Size and Mass

The diameter of the Sun was first determined by a sequence of indirect measurements and deductions. The Sun subtends an angle of about half a degree in the sky. If a thin triangle is drawn with an opening angle of half a degree, no matter how long the triangle, one finds that the short side of the triangle is always about 1 percent of the long side. Therefore, the diameter of the Sun is about 1 percent of the distance between the Earth and the Sun. The distance to the Sun was first deduced accurately in 1672, using a measured distance to Mars and Kepler's laws of planetary motion, which give the ratio of the distances Earth to Mars / Earth to Sun. The distance to Mars, in turn, was measured by triangulation between widely separated observers on Earth. The modern value for the average distance to the Sun, called the Astronomical Unit, is 150 million kilometers (93 million miles). Therefore, the diameter of the Sun is about 1.4 million kilometers (0.9 million miles).

The Sun is 110 times larger than the Earth. It is large compared to any of the planets, especially the Earth, which is one of the smaller planets.

In the seventeenth century Newton devised a method of estimating the Sun's mass. Newton realized that the Sun's gravity keeps the Earth moving around the Sun, rather than flying off into space at a high speed, in much the same way that the Earth's gravity keeps the Moon moving around the Earth. Once the Astronomical Unit was known, Newton's law of gravitation could be used to find the ratio of the Sun's mass to the Earth's mass. That value turns out to be quite large, a ratio of about 0.3 million. The Sun's mass is very large in comparison with the mass of any of the planets. The Earth's mass is truly inconsequential in the grand scheme of the solar system. Nevertheless, the Earth's mass causes the Earth's gravity and is therefore important for us living on the surface of the Earth.

Newton also made a remarkably good estimate of the mass of the Earth. He estimated that the mean density of the materials within the Earth was about 5.5 times that of water. The Earth's mass is the product of that density and its already known volume. From Newton's estimate of the Earth's mass and the ratio of 0.3 million follows the mass of the Sun.

The numerical value of the Sun's mass is very large. Saying that it is 0.3 million times the Earth's mass does not help much. Saying that it is a billion billion times the mass of the Himalayas also tends to boggle the mind. Perhaps it is most easily comprehended by reference to the future fate of the Sun. Somehow, the solar mass must provide the fuel for all the energy emanating from the Sun. Ultimately, the Sun must change. When might that occur? This lifetime may provide us with a mental proxy for the Sun's mass.

The lifetime of the Sun depends not only on the amount of the fuel but also on the efficiency of the fuel and on the rate at which the Sun produces energy. The rate of energy production was the first factor to be measured.

Nineteenth Century: Power and Luminosity

Ancient civilizations knew how to capture and use solar energy. But it was not until the nineteenth century, with the growing interest in steam engines and thermodynamics, that solar energy was measured quantitatively. William Herschel first measured captured solar energy with a thermometer in about 1800. About 0.7 kilowatt of radiative energy impinges on any area of 1 square meter facing the Sun. This is a prodigious amount of energy. The energy falling on merely 1 square kilometer is equivalent to a 700-megawatt (million-watt) power station. Today such a a power station is considered large but not unusual. This indicates the potential attractiveness of solar power: In principle, it is nearly unlimited compared with the power available from fossil fuels. However, the technology needed to harness the solar energy is still under development (see figure 1.1).

A further difficulty had to be overcome before one could evaluate solar energy production. The energy reaching the ground is only a fraction of all the energy reaching the top of the atmosphere. This fraction had to be estimated. It turns out that only about half of the arriving solar energy reaches the ground. The rest is either directly reflected back into space or absorbed in the atmosphere and later reradiated into space. The value now accepted for the average solar energy arriving at the top of the atmosphere is 1.4 kilowatt per square meter (or 2.0 calories per minute per square centimeter).

The Earth obstructs only a tiny fraction of all the solar radiation. The same amount of energy passes through all parts of a sphere that has a radius of one Astronomical Unit and is centered on the Sun. The total power leaving the Sun is called the solar luminosity. We could express the luminosity in terms of watts, or perhaps megawatts. But even the unit of megawatts is far too small a unit: Expressed in megawatts, the solar luminosity is a number that mere humans cannot comprehend. We need an example that involves numbers more nearly similar to those related to daily experience.

Imagine a bridge of ice stretching from the Sun to the Earth, 1 mile wide and 1 mile thick. We know that it takes energy to melt

Figure 1.1. Solar power generation. The "Power Tower" in Albuquerque, New Mexico, collects the energy reflected by 72 mirrors and uses it to create steam and electricity. (Sandia National Laboratories)

ice. Imagine now that we could use the energy from a 700-megawatt power station to melt this ice. How much time would be needed to melt all that ice? This is a prodigious amount of ice, and most people estimate that to melt it would take many years. Indeed, the 700-megawatt power station would take about 14 billion years to melt the ice. This is three times the age of the Earth and close to the age of the Universe. Yet the Sun would melt all that ice in merely a few seconds! This gives us at least some notion of the power emerging from the Sun.

Fuels: Wood or Coal?

How long could the Sun maintain its luminosity? That depends on the quality of the fuel producing the energy.

Plate 1. Skylab in orbit during 1973. The Apollo Telescope Mount and its four solar power panels face the Sun. X-ray and other cameras are mounted on the circular disk. One of the essential functions of the astronauts was to exchange film canisters from external cameras during extravehicular activities. Limited remote-controlled operation of the cameras continued between manned missions. (National Aeronautics and Space Administration)

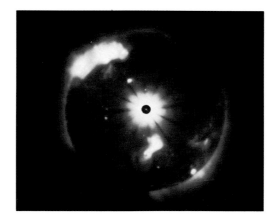

Plate 2. X-ray image of solar flare. The X rays, detected aboard Skylab, are converted into a false-color image by computer. The flare on August 6, 1973, occurred close to disk center. This composite shows a short exposure of the flare (inset) and a longer exposure of the solar X-ray image minutes before the flare. The flare is tiny compared to the Sun, but its diameter is comparable to that of the Earth. (National Aeronautics and Space Administration)

Plate 3. Eclipse of June 30, 1973. The eerie glow of the solar corona is visible only if the much brighter disk is totally obscured. (Jay M. Pasachoff, Williams College–Hopkins Observatory)

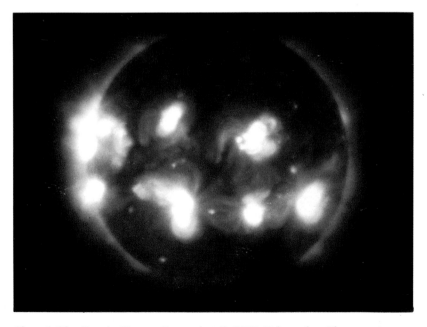

Plate 4. The Sun in X rays, September 5, 1973. False color. The most intense sources of X rays, appearing white, hover over large groups of sunspots. X-ray-emitting loops connect many portions of the corona. (National Aeronautics and Space Administration)

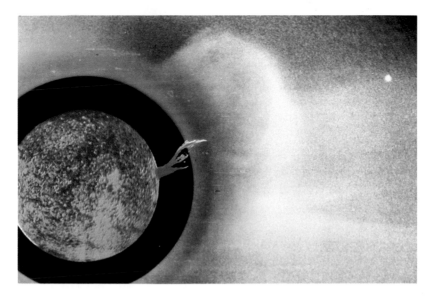

Plate 5. Transient of August 21, 1973. The bubble of coronal gas on the right has expanded to a diameter of almost three solar diameters. Such "transients" tend to expand, lift off the Sun, and fade into space in merely one to three hours. An occulting disk blocks the innermost corona (black). The image of the erupting prominence, observed at the beginning of the transient, is superposed. (National Aeronautics and Space Administration)

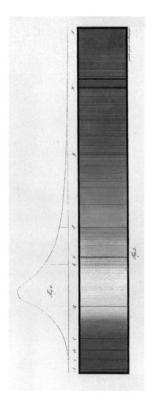

Plate 6. Fraunhofer's solar spectrum. Fraunhofer improved on the optical system of a simple prism. He found that the spectrum contained dark lines at certain specific colors. He labeled the most important ones by letters. The designations D and H are still in common use. The lines are now called Fraunhofer lines. They are described more accurately by their wavelength. Wavelengths included in this spectrum range from about 4000 Å (Angstroms) in the blue to about 6600 Å in the red. (Courtesy Deutsches Museum, Munich)

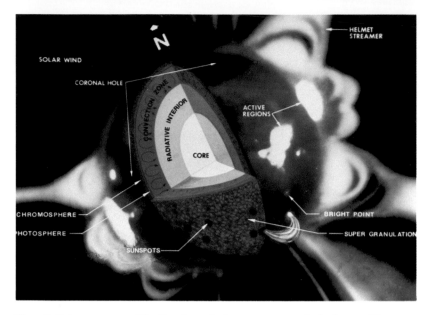

Plate 7. Solar structure. The Sun is entirely gaseous, mostly hydrogen. The core temperature is about 15 million degrees, which is hot enough for proton-proton fusion to produce the solar energy. The layers further from the center are progressively less hot and less dense. Each layer is adjusted so that gas pressure supports it against gravity and so that all the solar energy can diffuse out without piling up anywhere. The observable outer layers—that is, the solar atmosphere and its many features—are the subject of most of this book. (Courtesy Marshall Space Flight Center, National Aeronautics and Space Administration)

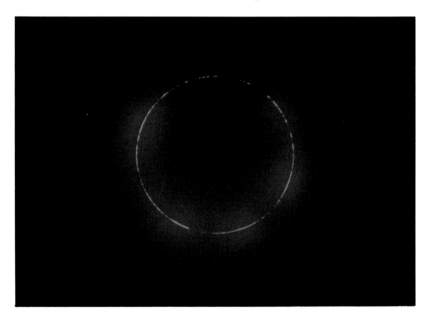

Plate 8. Annular eclipse of March 29, 1987. The Sun is covered only by the highest mountains on the Moon. (Photo by Huguette Guertin with an 89-mm Maksutov-Cassegrain lens at f/11, ASA 100 35-mm Kodacolor VRG film, and a 1/15-second exposure; copyright Huguette Guertin)

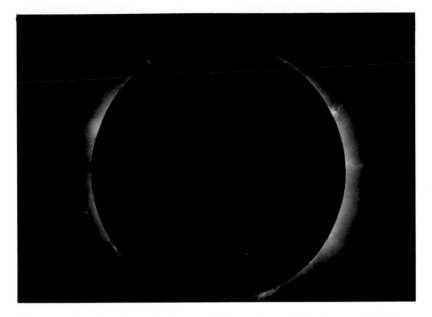

Plate 9. Total solar eclipse of February 26, 1979, observed in Brandon, Manitoba, Canada. The red prominences are part of the chromosphere, visible for only a few seconds during eclipse. The corona, in the shape of streamers or "helmets," tends to be brightest over the prominences. (Jay M. Pasachoff, Williams College–Hopkins Observatory)

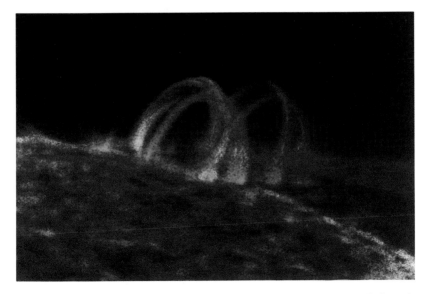

Plate 10. "McDonald's" Arches. This simple prominence was observed aboard Skylab on August 14, 1973, in the light of neon at 465 Å, representing temperatures in the transition region. (National Aeronautics and Space Administration)

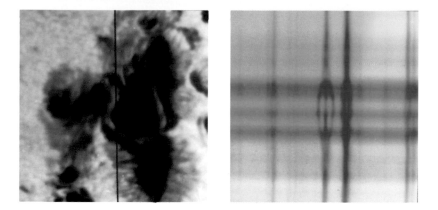

Plate 11. Zeeman splitting. At the left is a part of the solar image containing a sunspot, taken July 4, 1974. The light selected to reach the spectrograph passes through a slit shown as a black line. The resulting spectrum is shown on the right. Each vertical line is a Fraunhofer line. Two of them are widened and even split where the slit crosses the sunspot, an effect of the spot's strong magnetic field. Where the slit crosses the darkest part of the sunspot, the spectrum is darker at all wavelengths, not just at the Fraunhofer line. (National Optical Astronomy Observatories)

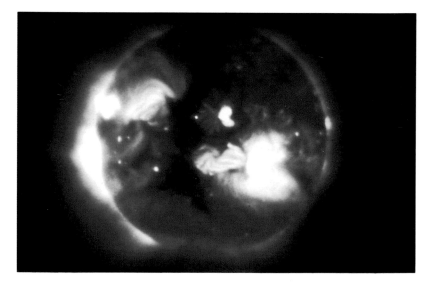

Plate 12. Coronal hole "number one" observed aboard Skylab in soft X rays on August 21, 1973. The hole appears in its most pronounced geometry, extending from one pole all the way past the equator. Ordinary differential rotation would have distorted the hole drastically during the next three months, yet the hole remained evident for at least six rotations. The hole was gradually filled by encroaching active regions. (National Aeronautics and Space Administration)

Plate 13. Clark Lake Radio Observatory, operated by the University of Maryland on a (usually) dry flat lake bed in California. It consisted of 720 spiral antennas, spanning 3 kilometers and tunable instantly to any of 1,024 frequencies in the range 15 to 125 megahertz. The Sun can be mapped with a resolution of a tenth of the solar diameter at the highest frequencies. (Courtesy M. Kundu, University of Maryland)

Plate 14. Aurora, photographed aboard Space Shuttle *Challenger* on flight 51B, launched April 29, 1985. The thin curtains occur at the "feet" of magnetic field lines whose upper portions reach the front of the magnetosphere (figure 10.5). (National Aeronautics and Space Administration)

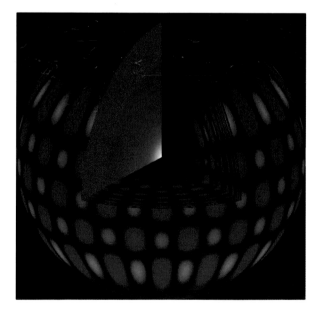

Plate 15. Patterns of motion throughout the Sun associated with one possible mode, or tone of solar oscillations. Blue implies radially outward motions, red inward. The directions of all the motions reverse after about 2.5 minutes, hence the name five-minute oscillations. The oscillations are sound waves that "fit" onto the solar surface, as in a bell, and also fit in depth, as in an organ pipe or pulsating stars. (National Optical Astronomy Observatories)

Up to the early nineteenth century, it would not have been unreasonable to suppose that the Sun is made of wood or coal. We now know that such a fuel would not last long, merely a few thousand years at the present rate of energy consumption. Of course, the solar luminosity would probably not be constant as the Sun gradually used up its fuel. More likely, the Sun would slowly dim. Moreover, it would have to have been brighter a mere few thousand years ago. For instance, at the peak of Egyptian civilization, it would have to have been about twice as luminous as now. The Earth would have absorbed very much more heat. It would have been so hot that life would have been impossible, but Egyptian historical evidence shows an abundance of life. Therefore, solar energy cannot be derived from wood or coal. Today we have many more arguments to disprove this hypothesis.

A clue that the Sun is not solid, nor even liquid, was discovered well over a century ago. The clue is contained in the value for the mean density of the Sun. That density is merely 1.5 times the density of liquid water and appreciably smaller than the mean density of the Earth. The Earth is dense because, within it, materials are highly compressed by the pressure of the overlying layers. This pressure prevents the Earth from collapsing under its own gravity. Because the Sun is so much more massive, the gravity within the Sun is much stronger than that within the Earth. Pressures in the Sun must therefore be much greater. Any solids within the Sun would be compressed to densities greater than those of the Earth. The low mean density of the Sun tells us that is not the case; it is a clue that the Sun is entirely gaseous and very hot on the inside.

Fraunhofer and the Solar Spectrum

Joseph von Fraunhofer (1787–1826) discovered direct evidence to show that the Sun must be gaseous, at least at its surface, through his work on optics. Among other projects, he worked on a more accurate method to disperse light into its various colors. Newton had already done that. Newton had passed sunlight through a pinhole and then through a prism and found that the various colors emerged in different directions. William Wollaston and soon thereafter Fraunhofer improved upon this method substantially. In particular, they let the sunlight fall through a slit (figure 1.2). Plate 6 shows Fraunhofer's result, the "spectrum" of the Sun.

Fraunhofer improved on the optical system to the point where he could map 576 dark lines. Clearly, color is too vague to describe such a map, and a system of letters suffices only for the most important lines. A numerical scheme is more useful. The most usual scheme is based on the wavelength of the light, expressed in Angstroms. One Angstrom (abbreviated as Å) is very small: 10 billion Å fit into 1 meter. Fortunately, we do not have to visulize 1 Å, but merely use it as a convenient scale for the wavelength and color of light. The wavelengths of the various colors can be read from the scale on Plate 6. They range from about 4000 Å in the blue to about 7000 Å in the red. One can therefore speak of a dark line at 6563 Å, or at 3933 Å. These figures are sufficiently accurate for practical use.

Fraunhofer's investigation showed the spectrum lines to be narrow and sharp. The narrowness of the dark lines alone guarantees that the solar surface is gaseous, because no solid or liquid yields a spectrum as simple as this.

Solar Chemical Composition

The further significance of Fraunhofer's lines became clear only later in the nineteenth century, when it was recognized that certain patterns of dark lines were characteristic of specific elements. Of special interest is a sequence of lines that seem to have a simple pattern in their spacing, which becomes narrower toward the blue. A few of these lines are labeled in plate 6. This particular sequence is characteristic of hydrogen. It indicates that hydrogen is present on the solar surface. The hydrogen is gaseous. The strongest of the lines due to hydrogen is at 6563 Å. It is usually referred to as the hydrogen alpha line.

In Fraunhofer's time, the understanding of atoms was insufficient to deduce just how much hydrogen is present on the Sun. Now we know that the Sun consists largely of hydrogen. Any

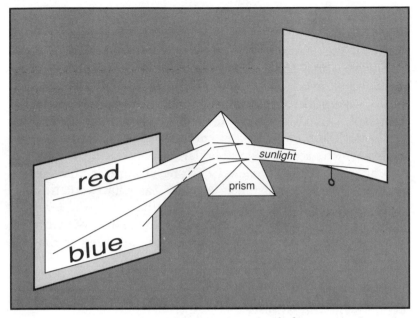

Figure 1.2. Spectrum formed by a prism. Sunlight arriving from the right passes through a slit under a window shade. It falls onto a prism, passes through, and falls onto a screen. The prism separates the incident light according to its colors. If only blue light were arriving, it would fall only onto one part of the screen and form a blue line there, a blue image of the slit. Red light would fall onto another part. "White" sunlight, being composed of all colors, is dispersed into a rainbow pattern. Newton already observed this kind of pattern.

part of the Sun, analyzed by mass, would show that about 75 percent of the material is hydrogen.

The next most abundant element is helium, about 23 percent of the solar mass. Helium also has a characteristic spectrum. In fact, when the periodic table of the elements had been largely filled in, an element was predicted in the position we now associate with helium. A spectrum of this element on the Sun provided the first proof that such an element exists. Hence it was named helium, from the Greek *helios*, meaning sun. However, no obvious spectrum lines of helium are visible in plate 6.

Almost all the other known elements also exist in the Sun, but they make up only about 2 percent of the solar mass. Of these, the most abundant are carbon, nitrogen, oxygen, magnesium, aluminum, silicon, and iron. In fact, all known stable elements exist in the Sun. All are gaseous. The abundances are typical of most matter in the Universe. Abundances on planets like the Earth are the exception.

The dark lines of the solar spectrum are caused by atoms on the Sun that prevent the light at these wavelengths from reaching us. One speaks of the atoms absorbing the light. Accordingly, Fraunhofer's dark lines are referred to as absorption lines.

Most of the prominent absorption lines are due to hydrogen atoms, which are the most abundant. But some of the atoms of low abundance also have very strong absorption lines in the solar spectrum. For instance, calcium has what we call the H and K lines at 3968 Å and 3933 Å, respectively, shown in plate 6. These lines are even more prominent than the lines of hydrogen. Their strength is due to atomic details, not to a high abundance of calcium.

Lord Kelvin and Gravity

Wood and coal yield energy from chemical reactions. Once they were eliminated as a possible source of solar energy, Lord Kelvin (William Thomson, 1824–1907) considered a different form of energy, namely gravity, a likely candidate. When we drop an object, it speeds up under the influence of gravity and gains kinetic energy, which is turned into heat upon impact. Gravity can also provide energy in a more gradual manner. Suppose that the Sun shrinks slowly, under the influence of its own gravity. As the gases become compressed, they tend to heat up, much as the air heats up in an inner tube being inflated. The heat can migrate outward toward the surface of the Sun, and from there it escapes as radiation. Might the Sun shine because it is shrinking?

Lord Kelvin asked himself how long the Sun could last if it were gradually contracting. He estimated twenty million years. At the time, twenty million years seemed enormously long as geologists thought they measured the age of the Earth in mere thousands of years. In fact, Lord Kelvin pursued this subject because he wanted to show that the Earth is at least several million years old. Solar contraction lasting several million years fit naturally into this picture. Therefore, Lord Kelvin viewed gravitational energy as an attractive explanation for solar luminosity.

Kelvin's explanation for solar luminosity implies that the Sun should shrink continually. Its luminosity should weaken with time. The energy flux arriving at Earth a hundred or more million years ago would have to have been substantially larger than today, and the Earth much hotter. Yet today we have evidence of primitive life dating back three billion years ago. Liquid water must have existed even then. The Sun cannot have been shrinking at the rate Kelvin supposed during all this time. Therefore, gravity is not the present source of solar luminosity.

According to modern radioactive measurements, the Earth is 4.6 billion years old. All evidence for the formation of the planets suggests that the Sun is at least equally old. The Sun has been remarkably constant during all this time. A source of energy must be found that is much more efficient than either chemical reactions or gravity. Today we know that it is nuclear energy.

Einstein and Nuclear Energy

Nuclear energy is at the heart of Einstein's famous equation, $E = mc^2$, formulated in 1905 in connection with his special theory of relativity. The equation is best interpreted as: mass may be turned into energy, or vice versa. The energy that might be obtained by converting a single gram of matter into energy is enormous. If this energy were somehow liberated under the author of this book, to accelerate him upward, it would propel him not just to the height at which airplanes fly, not just out of the Earth's atmosphere, not even merely out of the gravitational domain of the Earth or the solar system, which would hardly slow him down, but it would propel him out of our Galaxy into intergalactic space.

The Sun turns mass into energy. We know how much energy it produces per second. Therefore, Einstein's equation tells us how much mass is destroyed per second. That turns out to be 4 million tons per second. The Sun has been on this "reducing diet" for 4.6 billion years, yet the total mass lost is a negligible fraction of the Sun's mass.

Einstein's equation does not specify how mass can be changed into energy. Certainly, normal matter cannot be totally converted into energy. In 1938, Hans Bethe found a way that mass could actually be changed into energy through a realistic sequence of nuclear processes. The net effect is that four hydrogen atoms are converted into a single helium nucleus. If a ton of hydrogen is converted into helium, the helium weighs about 0.007 ton less. The missing mass has been converted into energy. The solar net conversion of 4 tons of matter per second into energy requires the conversion of 500 million tons of hydrogen into helium per second. How long before all the solar hydrogen is used up at the present rate of conversion? The answer is about 100 billion years. Clearly, this poses no conflict with ages on Earth. It also shows that nuclear energy is a fantastically efficient source of energy compared to chemical reactions and gravity, lasting millions and thousands of times longer, respectively. This efficiency explains why there is great interest in creating industrial nuclear-fusion energy on Earth: The hydrogen in the oceans would provide a practically inexhaustible supply of fuel. The problem in creating this bonanza is that nuclear fusion requires extremely high temperatures. No fusion occurs at the ordinary temperatures we find on Earth.

Nuclear Fusion

The first stage of nuclear fusion is the hardest to bring about: Two hydrogen nuclei must collide, or at least pass close enough to each other so that the nuclear forces have a chance to act. They must pass within a trillionth of a centimeter of each other. That position is difficult to achieve because both nuclei are positively charged and therefore repel each other. If they approach each other slowly, they are deflected away and can never get close enough to the critical position. High speeds are essential for a reasonably probable collision that results in fusion. The necessary high speeds occur only in gases at high temperatures, in the range of 10 to 100 million degrees.

Even the physical collision of two protons does not ensure fusion. During the instant in which the protons are within a trillionth of a centimeter of each other, nuclear forces must convert the two protons into three new particles: namely, a deuteron, which is heavy hydrogen; a positron, which is a positively charged electron; and a neutrino, which is charge-neutral and escapes from the Sun (figure 1.3). This interaction involves the so-called weak nuclear forces. The adjective "weak" signifies that the chance for interaction during any one collision is extremely low.

Solar Structure

Normally on Earth, gases expand as they are heated. The million-degree gases produced in a nuclear bomb expand vigorously. Such hot gases are difficult to contain for laboratory fusion and fusion machines. But they are easily contained near the center of the Sun, because the weight of the overlying gases is enormous. The outward pressure of the hot interior gases is exactly balanced by the inward force of gravity.

The temperature at the Sun's center turns out to be about 15 million degrees. The matter there is entirely gaseous, most of it hydrogen. The electrons of all the atoms have been stripped from their nuclei. One must imagine a mixture of hydrogen nuclei, helium nuclei, a few other nuclei, and electrons all moving about each other at high speeds. Typically, they collide with each other many times per second, well before they have moved even 1 centimeter. Very, very rarely, two proton nuclei will collide so nearly head-on and with such unusually high speeds that they fuse and release energy. A typical proton at the solar center has only a 50 percent chance of having fused even after five billion years. Indeed, about half of all original hydrogen atoms at the solar center have been converted to helium.

Gases at 15 million degrees are filled with intense radiation. The radiation diffuses outward, toward less hot solar layers, in much the same way that sunlight diffuses through a fog on Earth toward darker regions. The radiation carries the energy with it. Ultimately, after about 20 million years, the energy reaches the solar surface, where it escapes into space as sunlight.

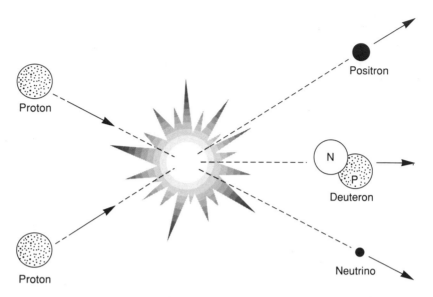

Figure 1.3. Proton-proton fusion. Sometimes, though very rarely, two protons approach each other so closely that nuclear forces change the two protons into three new particles. The deuteron is a heavy hydrogen nucleus, the positron is a positively charged electron, and the neutrino is a particle that escapes the Sun.

Plate 7 outlines the structure of the Sun. The temperature is highest near the center, where it measures about 15 million degrees; it gradually decreases to about 10 million degrees half-way out, then decreases ever more precipitously to a mere 5,000 degrees at the surface. The density is also highest at the center, about 150 times the density of water on Earth. But the pressure, a trillion times atmospheric pressure at sea level on Earth, crushes all atoms and ensures that all matter remains gaseous. The density decreases outward quite gradually until about a tenth of the way out, but further out it decreases rapidly until it reaches a value at the surface that is merely a millionth of the Earth's atmospheric density at sea level.

These values are derived from theory. In principle, the theory satisfies fairly simple requirements. The mathematical model for each gaseous layer of the Sun is adjusted so that the layer is supported against gravity by the layers below it and so that it, in turn, can hold up the layers above it. It is also adjusted so that all the radiation created near the center can migrate outward without piling up anywhere. If we knew all about nuclear reactions and the manner in which energy travels outward, we could predict a unique value of the solar luminosity and radius, given only the solar mass and chemical composition. In fact, the gases in the outer third of the Sun are turbulent. These motions help to convect the heat upward. (Hence there exists the region marked convection zone in plate 7). Convection is not well understood. As a result, the theory of solar structure is not quite complete. The deficiency is considered worrisome but probably unimportant by most astronomers.

The solar "surface," the layer from which we obtain our sunlight, appears to have a sharp edge, as normally seen from Earth. In fact, modern observations show it to be an extended gaseous layer. The source of sunlight, referred to as the photosphere, is just one layer of a more extensive "solar atmosphere." The atmosphere includes the sunspots, the solar flares, the corona, and many other phenomena, which are the subject of most of this book.

Solar Lifetime

If the escaping radiation were not replenished near the center, most of the radiation present in the Sun would escape in about twenty million years. This would cause the Sun to shrink gradually and to change its size and luminosity significantly in some twenty million years. Indeed, the time needed for much of the radiation to escape is the same as the time Kelvin estimated for the Sun's lifetime. Kelvin was correct in his calculation of the rate at which the Sun loses energy to space, but he could not know that this energy is replenished steadily by nuclear reactions.

How long can the nuclear "fires" last? Only the central parts of the Sun are hot enough to yield nuclear fusion. Something drastic must happen when the central hydrogen is exhausted. Detailed computations indicate that the central hydrogen will be exhausted after about 10 billion years, when the Sun has used up the central tenth of all its hydrogen. Ten billion years, give or take a billion, is considered the life span of the Sun.

The Sun is now about halfway through its life span. We do not yet have to worry about future changes. However, we should appreciate the constancy of the Sun over the past four billion years. The solar constancy, together with some evolution of the Earth's atmosphere, has ensured the presence of liquid water on Earth throughout this time. Without liquid water, life would not have reached its present state.

Solar Fate

The Sun must change when the hydrogen near the center is exhausted, which will be in about five billion years. The Sun will grow to several times its present size. Its radiative power will increase to over a hundred times the present value, fueled by the fusion of three helium nuclei into carbon. Although the Sun will be somewhat less hot at the surface than it is now, it will loom like a large furnace in the sky as what the astronomers call a red giant star. Its effect on the planets will be devastating. Mercury will be enveloped and will evaporate. Venus will orbit just beyond the solar surface. Earth will become scorched and lifeless. Only Pluto might be a suitable abode for any life remaining in the solar system.

The nuclear fusion that powers the Sun while it is a red giant will last only about 100 million years. When the Sun again runs out of nuclear fuel, it will blow off its outermost layers. What is left will shrink to the size of the Earth. At first, the shrinking and compression will release energy, so that the Sun will be a white-hot object, what is called a white dwarf. It will stop shrinking when it reaches the size of the Earth and thereafter will cool off very gradually. It will once again become a red star, but now will be the size of the Earth, after some 20 or 30 billion more years, if the Universe lasts that long. This stage constitutes the stellar graveyard. Any remaining planets will freeze thoroughly. If any life exists on Pluto when the Sun becomes a white dwarf, it will have to migrate to some planet near another star.

Chapter 2

Solar Interior: Neutrinos

Star Tests

Most observations of the Sun pertain to its outermost layers, its atmosphere. The thickness of the layer of gas in which we see the sunspots is less than 1/1000 of the solar radius. We almost literally see only "skin-deep." Most of our information about the solar interior is derived from theory, reinforced by observations and theoretical modeling of other stars. The theory has been remarkably successful in describing the kinds of stars observed.

Every star goes through the same two stages of the stellar life cycle that the Sun does: a long stage in which hydrogen is the nuclear fuel, as in the Sun now, and a brief stage in which helium is the nuclear fuel and the star is a red giant. We observe the stars during a few decades, which amount to no more than a fleeting moment in the life of a star. Most stars are found in the hydrogen-burning stage. Not surprisingly, fewer stars are found in the relatively brief red giant stage. The numbers of stars found in various stages of their life cycles are very close to theoretical predictions. Therefore, astronomers have great confidence that stellar theory is basically correct. Nevertheless, there are some nagging uncertainties in stellar theory, notably with respect to convection. These uncertainties loom larger as observations rapidly become more precise. Therefore, any method that enables us to "look inside" the Sun improves our understanding of both the Sun and all the other stars.

The Sun is near and bright enough to "look inside" with the aid of two techniques that are still too difficult to use on other stars. One technique is to measure "solar oscillations." It turns out that the Sun "rings" in much the same way that a bell rings. We can learn even more from these "tones" than we can learn by listening to a bell. As the newest and still rapidly developing field of solar research, it is best discussed near the end of this book (see chapter 16). The other technique is to measure particles called neutrinos.

Neutrinos

Neutrinos have a most unusual property: They pass through all matter so easily that, once formed at the solar center, they can freely traverse the Sun, emerge from it, and continue on into space. We want to measure these particles. To do so, we must try to capture them in our relatively tiny, Earth-sized instruments after they have has passed uncaptured through all the material of the Sun. Clearly this is a difficult feat!

Fermi first predicted the neutrino in 1933: He noted that energy and momentum appear to be lost whenever a neutron decays into a proton. He argued that the apparently lost energy and momentum are actually carried away by an unobserved particle. That particle is the neutrino. Neutrinos were first detected in the early 1950s. Detectors weighing many tons were placed next to one of the most powerful nuclear reactors available, and the scientists proudly reported that they had detected a few dozen neutrinos! To measure solar neutrinos, we require one of the most precise nuclear detectors ever designed. One such detector has been running for almost two decades.

Neutrinos are an essential by-product of the fusion of two protons (figure 1.3). That fusion is only the first of a series of reactions that convert hydrogen into helium (figure 2.1). The first step of fusion is also shown in figure 1.3. When two protons fuse, a deuteron (proton plus neutron), a positron, and a neutrino emerge. Almost immediately, the positron merges with an electron to become radiation, in the form of a gamma ray, and this radiation is almost immediately turned into heat, which provides part of the solar luminosity. At the same time, the neutrino escapes.

Proton fusion is the step that occurs least readily in the se-

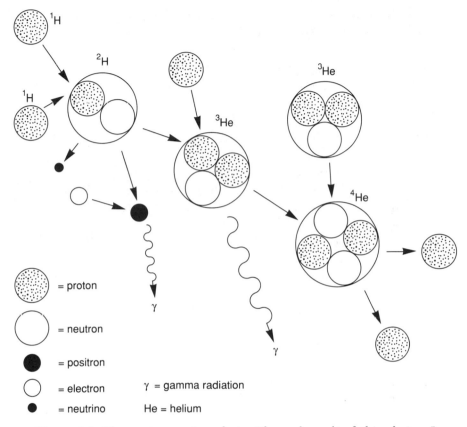

= proton

= neutron

= positron

= electron   γ = gamma radiation

= neutrino   He = helium

Figure 2.1. The proton-proton chain. The end result of this chain of nuclear reactions is the conversion of four protons and two electrons into a helium nucleus and a neutrino. The helium nucleus weighs less than the initial particles. The difference appears as energy, namely heat and the neutrino.

quence of reactions. The next steps occur relatively easily and quickly. At this point, the deuteron merges with a proton. That means two protons are now present in the new nucleus, which signifies that helium is being formed, but the single neutron makes it merely a light isotope of helium. A gamma ray is emitted that is promptly turned into heat. Finally, most of the time, two of these light isotopes of helium merge to form a single normal helium nucleus, liberating two protons. After each reaction, the protons, deuterons, or helium nuclei are moving at high speeds and their kinetic energy is promptly turned into heat. The end result is that four protons and two electrons are converted into a helium nucleus, heat, and a neutrino.

How can one possibly capture something that has passed all the way through the Sun? A typical neutrino easily passes through the Earth, and certainly through the substance of any nuclear detector. Fortunately, the number of solar neutrinos passing through us is enormous. As I write this sentence, roughly 100 trillion neutrinos are passing through me.

Although trillions of neutrinos pass through me each second, I am, astronomically speaking, so tiny that I capture at most a neutrino per day, something that I will never notice. I make a rather poor detector of neutrinos, because at the same time I am capturing many more other nuclear particles that are created naturally on Earth. One more nuclear reaction within me makes no difference.

When a neutrino is captured by some nucleus, it alters the nucleus or even destroys it. Captures that make the nucleus radioactive are of special interest, for then we may attempt to detect the decay of that radioactive nucleus. If we can show that only a neutrino could have caused an observed radioactive decay, then we have in effect measured a neutrino.

To detect solar neutrinos we require a material that responds only to neutrinos, and to nothing else. Moreover, we need a large detector to capture even a few neutrinos.

Some 20 years ago, Ray Davis devised a detector made of cleaning fluid (perchloroethelene). His most recent version contains 100,000 gallons of fluid (figure 2.2). The material is cheap enough to obtain in large quantity. Most important, when a neutrino is captured by one of the chlorine nuclei, a radioactive argon nucleus is created. The radioactive argon nuclei can be chemically removed from the cleaning fluid and then counted. Few other nuclear reactions would create radioactive argon. Therefore, the presence of the radioactive argon is a clear sign of captured solar neutrinos.

Although the chlorine in cleaning fluid is cheap enough to obtain in quantity, it has one major disadvantage. It readily captures only a small fraction of the solar neutrinos; that is, it

Figure 2.2. The neutrino experiment at the Homestake Gold Mine, Lead, South Dakota. The tank contains 100,000 gallons of cleaning fluid. When a chlorine nucleus captures a neutrino, it turns into an observable radioactive nucleus of argon. (Brookhaven National Laboratory)

captures only the most energetic ones. These are not emitted during the sequence of reactions in figure 2.1, but only when fusion proceeds through one of a few relatively rare alternate sequences. The most important is a sequence via beryllium, shown in figure 2.3. Only a small fraction of all fusions go through this branch, and the chlorine detects or captures only the most energetic of the resulting neutrinos. Therefore, the capture rate is very low indeed.

Davis's detector is designed for and was expected to measure roughly one radioactive argon atom per day. The sensitivity is astounding. One atom per day is to be counted after it is created in a vat containing 100,000 gallons of matter! One might be tempted to wait a few months, or even years, in order to let the argon accumulate. Unfortunately, that is not possible because the radioactive argon atoms decay. Half of them are gone after 35 days, whether they were measured or not. At most, about 60 radioactive argon atoms were expected to exist in the tank at any one time. Davis's experimental setup is designed to extract these 60 argon atoms about once a month. After each extraction, the argon is transferred to a counter. It is then necessary to wait for the radioactive nuclei to decay and to measure the number of these decays.

Davis placed his detector in the Homestake Gold Mine at Lead, South Dakota. Few cosmic rays reach that far below the ground, so that the background of unwanted nuclear reactions is low. Davis also installed electronic detectors in his counters so that he could distinguish between the desired and the background counts. The materials that he selected for the counters were such that the residual radioactivity of the counters would not exceed the other background signals. In this fashion, Davis reduced the counts from cosmic rays and natural radioactivity, the "background rate," to a mere 10 counts per month. This is an impressive technical achievement. This background rate in itself has been measured only very recently.

Surprise

The first results were obtained in the early 1970s. They were rather embarrassing. The observed rate of neutrinos was only

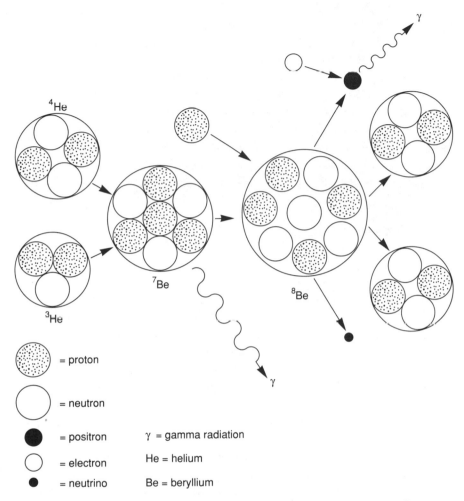

Figure 2.3. The beryllium branch of the fusion chain. Occasionally, a light helium nucleus fuses with a normal helium nucleus. A neutrino is emitted in the subsequent reactions. Davis's experiment can detect only the more energetic of these neutrinos, that is, only about one in 10,000 solar neutrinos.

about a fourth of the expected rate! Was the experiment less accurate than claimed? Were some of the argon atoms missed? Many people suggested possible causes for errors. All possibilities were checked out carefully and eliminated. The measurements were repeated many times. During the next few years, the observed rates averaged slightly higher, but they did not reach the levels expected. Today, the best average rate obtained is about one-third that expected.

This result is particularly frustrating to astronomers because the theory for stellar interiors works so well for astronomical observations. Therefore, astronomers tend to look for an explanation in physics. However, many physicists tend to look for an explanation in astronomy. The author prefers an explanation in physics. The following summary of possible explanations may reflect that preference.

Solar Central Temperature

Time and ingenuity have produced many possible astronomical explanations of the discrepancy. All these explanations focus on the fact that the captured neutrinos are the most energetic ones. Only about 1 out of 10,000 falls in this category. The frequency with which fusion proceeds by this relatively rare pathway strongly depends on temperature. All theories predict a central temperature of "about" 15 million degrees, some more, some less. Any theory that reduces the predicted temperature at the center of the Sun by a rather small amount, while still accounting for the solar luminosity, reduces the predicted neutrino capture rate by a large amount.

Here are four suggestions that are generally treated with considerable skepticism: (a) The solar interior might fluctuate on a time scale of tens to hundreds of millions of years. Since radiation takes some 20 million years to emerge from the Sun, the present luminosity represents the nuclear reactions at the solar center some 20 million years ago. But the neutrinos emerge instantly and represent the nuclear reactions occurring now. If the interior temperature and the rate of fusion are slowly decreasing, then we are observing fewer neutrinos than indicated by the present luminosity. This theory implies that the Earth is subject to major long-term climatic changes. They might be related to periods of major ice ages separated by roughly 100 million years. (b) Rapid rotation near the center might provide some support against gravity, allowing a lower central pressure and temperature. (c) The Sun might have formed with fewer heavy elements like iron near the center than the outside. Since the heavy elements are the main absorbers of radiation, smaller amounts of them make it easier for the radiation to pass out of the center and help maintain a lower central temperature. However, another possibility is that substantially higher amounts of heavy elements are present. In that case, the radiation would drive convection near the center, the convection would transport extra hydrogen toward the center, and the additional nuclear fuel would allow the present luminosity to be produced at a lower temperature. Such inhomogeneous stars are anathema to most astronomers, since stars are thought to be very well mixed internally during the process of formation and no subsequent segregation of elements deep within a star has been demonstrated. (d) Some newly predicted particles called WIMP (weakly interacting massive particles) may help carry energy outward from the center. This would also produce a lower central temperature.

The measurement of solar oscillations may soon provide a test to indicate whether the solar interior rotates rapidly (preliminary results suggest that it does not) and may later provide a direct measure of the central temperature.

The neutrino data suggest one other possible explanation. Figure 2.4 shows the observed count rates for each measurement. The vertical error bars indicate the statistical uncertainties that are inevitable when one measures only a few argon decays at any given time. The observed rates were particularly low during 1970, the first year of measurements, and during 1980, when two measurements yielded merely the background rate. These two years happen to be the years with the highest numbers of sunspots. Is it possible that whatever causes the sunspots also suppresses the neutrinos? No theory even remotely suggests this. Probably the observed minima are merely fluctuations in the argon count rates. However, the first runs in 1986–87, after the experiment was shut down so that equipment could be replaced, showed relatively high neutrino fluxes, as might be expected for this time of near solar minimum if the solar cycle really influences the neutrino flux. It is certainly tempting to run the experiment for a few years further, just to check the results past the next solar maximum in 1991.

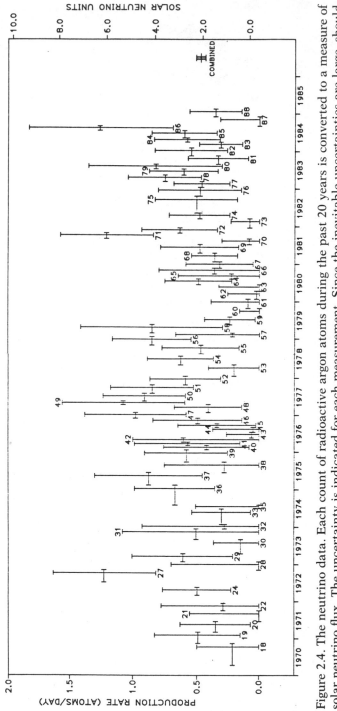

Figure 2.4. The neutrino data. Each count of radioactive argon atoms during the past 20 years is converted to a measure of solar neutrino flux. The uncertainty is indicated for each measurement. Since the inevitable uncertainties are large, should one consider only a time average as a reliable result? Or might the minima of 1970 and 1980, at times of sunspot maximum, be significant? (Raymond Davis, Jr., and Brookhaven National Laboratory)

Basic Physics

Is the neutrino deficiency the result of using incorrect physics in the neutrino predictions? The essential nuclear reactions in the solar interior cannot be checked in the laboratory. The first step in the conversion of hydrogen to helium, or the fusion of two protons, takes far too long to be detected in any appreciable amounts of nuclear fuel available on Earth. We cannot wait for the billions of years available for fusion at the solar center. Therefore, the prediction of this reaction rate must depend on nuclear theory and on the interpretation of other nuclear experiments. Conceivably, some physical processes have been omitted from the theory. Nuclear physicists have reviewed and remeasured the nuclear reaction rates used to predict the neutrino flux. The current consensus is that the nuclear reactions are not the problem.

Basic physics has provided one other possible way out of the dilemma. Recent theory "unifies" the nuclear forces (which cause nuclear fission), the weak forces (which cause radioactivity), and the electromagnetic forces (which cause emission of light). These forces and their effects are thought to become indistinguishable under incredibly hot conditions, much as various chemical processes do when conditions are hot enough to ionize all the atoms. The unified theory is being tested through its prediction of the extremely gradual decay of protons. So far, no protons have been observed to decay. However, this negative result has disproved only the simplest version of the theory and leaves many other versions untested. The theory is attractive to astronomers because it may help to explain several observations concerning the origin of the universe.

The unified theory also implies that the solar neutrinos can decay into three forms of neutrinos. Only the original form is captured by the chlorine in Davis's detector. If the solar neutrinos were to decay equally into all three forms of neutrinos, by the time they reach the Earth, then the observed rate should equal one-third of the originally predicted rate. The coincidence of prediction and observation is attractive. However, the manner and efficiency of this decay is still the subject of debate.

Do We See the Sun at Night?

Although the basic theory is several years old, one important point about solar neutrinos was recognized only in 1986. If some still unknown physical parameters, including the mass of the neutrino, happen to have suitable values, then the originally emitted form of neutrino is forced to change systematically to another form before it emerges from the Sun. This would explain why fewer neutrinos are observed than originally predicted. However, the ratio of one-third is then merely a numerical accident.

If the neutrinos are forced to change their form while passing out of the Sun, it is possible that they change also while passing through the Earth. The Earth may change some of the "wrong" form of neutrinos back to the observable form. This leads to the seemingly paradoxical suggestion that the present detector captures most of its neutrinos during the night, when the neutrinos must pass through the Earth and are partly converted back into the observable form.

Do we "see" the Sun via sunlight during the day and via neutrinos during the night? It may be possible to measure the radioactive argon atoms every half day. Then one can add up separately the results obtained during the daytimes and those produced during the nights. This means sampling all 100,000 gallons of fluid every 12 hours in order to determine whether a single radioactive argon atom has been produced during that time!

International Collaboration

The chlorine-containing fluid that Davis selected as a detector is readily available. Its drawback is that it measures only the relatively rare energetic neutrinos. Well over a decade ago, it was suggested that one could also build a neutrino experiment using gallium. Gallium captures about half of the neutrinos resulting from the fusion of two protons, plus many of the other neutrinos. The neutrino capture rate is about twenty times the originally predicted rate for chlorine. Such a measurement would complete and, we hope, corroborate the test for the solar interior. It would also shed light on the fate of the energetic neutrinos and the associated theory.

The problem with a gallium experiment is the cost of the gallium. No agency in the United States has been convinced of the need to spend some 100 million dollars for the gallium needed, even though it could be reused after a few years. However, two such experiments are being built elsewhere, one in Italy and one in the Soviet Union. The United States has the unusual opportunity to participate in the Russian experiment—because Davis and his collaborators have the technology to make the necessary pure counters—and to do so before the European community builds a competing experiment. The first gallium results may be available by about the time this book is published.

Chapter 3

Eclipse!

Totality

It is dangerous to look at the Sun with the naked eye except during a total eclipse. Our eye would focus the sunlight so that it would burn the retina. Fortunately, we instinctively turn our eyes away before that happens. Actually it is usually not very interesting to look at the Sun with the naked eye. On the rare occasions near sunset or sunrise when enough haze obscures the Sun so that it is safe to look at, there is little to see on the solar surface. Sunspots may be recognized only if they are unusually large. This happens only in periods of a particularly intense sunspot maximum.

Solar eclipses, which occur more frequently, provide a better opportunity to observe the Sun with the naked eye. At intervals of about six months, the Moon may pass partly or completely over the bright disk of the Sun. When it covers the Sun completely, one can see a faint outer glowing layer, which is impossible to see in the normal glare of the Sun: the solar corona.

By no means do total eclipses occur twice a year. Too many conditions must be satisfied. For one thing, the Moon must appear to be larger than the Sun. The apparent size of the Moon is always close to that of the Sun, but sometimes a bit larger, sometimes a bit smaller, depending on the changing distances of the Earth from the Moon and from the Sun. Conditions are favorable for a total eclipse if the Moon is nearer and the Sun is further than average. Second, the Moon's path must appear to cross the Sun, as seen from the Earth. Of course, the Moon always appears near the Sun during the lunar new phase. But most of the time, the Moon passes the Sun without even partly overlapping it. It must completely overlap the Sun for an eclipse to occur. These circumstances are relatively rare. They usually occur once every year or two.

Shadowplay

As the Moon moves past the Sun, its shadow sweeps a path across the Earth. People in the shadow path see an eclipse. The path for a total eclipse is typically less than 100 kilometers wide, although it may be many thousands of kilometers long. Some shadow paths appear in figure 3.1.

Even when fully visible, a total solar eclipse is a brief event, usually lasting less than four minutes and at most eight minutes. Amateur and professional alike can see a great deal in that brief time. Therefore, people travel long distances to view the eclipse in its widest zone. Most prefer to view the eclipse at its center, where it lasts longest. Unfortunately, nature provides little choice in these sites. Frequently, eclipses occur over the ocean or in deserts or jungles that are hard to reach.

Eight total eclipses will occur during the 1990s. None crosses the continental United States. An eclipse of unusually long duration, nearly seven minutes, will occur on July 11, 1991. With somewhat shorter durations, it will pass over Hawaii, Mexico, Central America, and Brazil. Its track is included in figure 3.1. More details of its path over Hawaii, where totality will last over four minutes, appear in figure 3.2. The island of Hawaii is unusually accessible and will be a popular eclipse site. Other eclipses lasting longer than three minutes will occur on June 30, 1992 (South Atlantic), November 3, 1994 (Chile, Brazil), and February 26, 1998 (Central America).

Safety

Someone watching a total eclipse need not be concerned about damage to the eyes as long as a few simple precautions are taken. The experience will live in memory for many years.

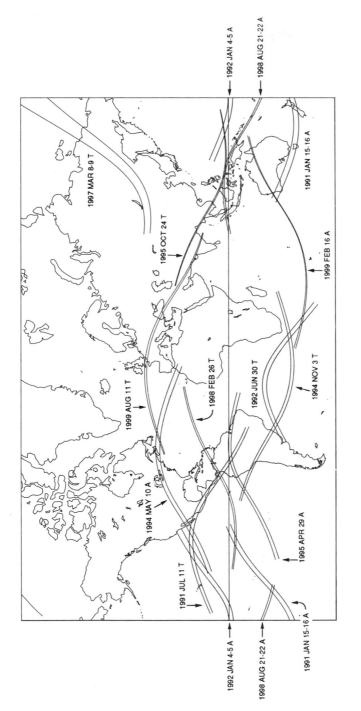

Figure 3.1. Eclipse paths 1991–2000. Any one within such a path observes a total (T) or annular (A) solar eclipse. The longest possible total eclipse seen from any one place lasts seven minutes (Mexico, July 11, 1991). Observers in an airplane following the eclipse path may see a slightly longer eclipse. Note: Eye protection is essential while even a tiny part of the Sun's bright disk is visible. Unprotected viewing is safe when, and only when, the bright disk is totally covered up. (Courtesy Nautical Almanac Office, U.S. Naval Observatory)

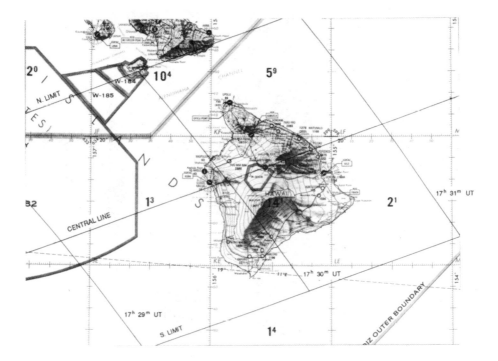

Figure 3.2. Path of eclipse over Hawaii, July 11, 1991. Observers on the central line on Hawaii will see the Sun eclipsed near the eastern horizon (altitude about 20 degrees) lasting 4 minutes and about 11 seconds. Totality can be observed anywhere between the northern and southern limits, but it is shorter further from the central line. The time of eclipse midpoint is indicated. (UT is Universal or Greenwich time; Hawaii time is 10 hours earlier.) (Courtesy Nautical Almanac Office, U.S. Naval Observatory)

The eclipse takes place in stages. Roughly two hours before totality, the Moon begins to cover up the Sun in a partial eclipse. The Moon, of course, shows us its shadowy side and appears dark and invisible, but it can be recognized by its outline on the part of the Sun that is covered up. Gradually, more of the Sun is "eaten" (figure 3.3). This is the time that is dangerous for observers, because the part of the Sun that remains uncovered is almost as intense as if there were no eclipse and can still burn the human eye. Even watching the Sun through a highly darkened

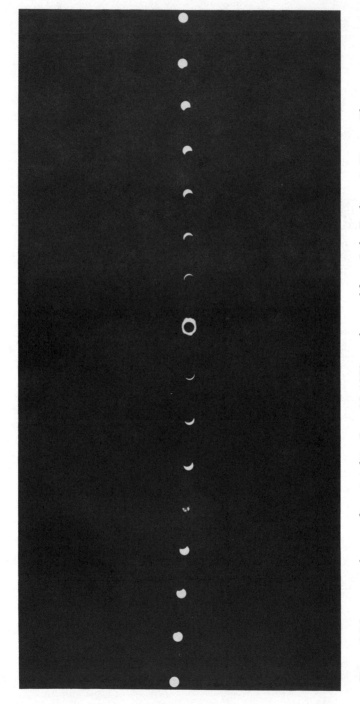

Figure 3.3 Ingress and egress from totality, June 30, 1973, as observed from Lake Turkana in Kenya. The exposures were made every 10 minutes. Clouds intervened at the time of the fifth image. (Courtesy Dale R. Corson)

glass or film is dangerous, because these items may transmit ultraviolet or infrared radiation that we do not feel; and because the Sun appears darkened, we do not involuntarily avert our eyes, but continue to focus the radiation, and thus may damage the eyes. The ultraviolet and infrared radiation is always present when the Sun shines, but normally we do not attempt to look at the Sun.

It is safest, and quite adequate, to project the Sun through a small hole in a cardboard onto a piece of paper. This simple device will make the "bite" taken from the Sun quite visible.

The eclipse becomes particularly interesting about 20 minutes before totality. The sky begins to darken. Birds may behave as they do in the evening and seek their roosts. Mosquitoes may also emerge. When I watched an eclipse in Alaska in 1963, my colleagues and I were warned well beforehand to apply mosquito repellant even though there were no mosquitoes around at the time. Sure enough, the people who ignored this good advice were hunted by mosquitoes throughout the period of darkness and were too distracted to observe the eclipse they came to see. The temperature also drops. If you are not in the tropics, put a sweater on your children before the eclipse begins, while you have time.

Lunar Mountains

The Moon is not quite round. Its mountains are about the same height as the mountains on Earth. They become distinct during the last few seconds that the solar disk is visible. Plate 8 shows a spectacular photograph of an annular eclipse: The Moon covered the Sun only with its highest mountains. Annular eclipses have been ignored somewhat by astronomers, but they are now gaining attention because some of them yield a precise measure of the solar diameter. An annular eclipse will cross the United States on May 10, 1994.

In the instant before a total eclipse, the solar disk is still visible through the deepest valleys on the Moon. The spark of light barely visible through these valleys is called the diamond ring effect. It is dazzling. It is terribly tempting to watch for it with the naked eye. DON'T! Even if the eye is not damaged permanently, it will become temporarily blind and the main phase of the eclipse will be missed. The diamond ring should be saved until the end of the eclipse.

Chromosphere

As the solar disk disappears, the Moon seems to become enveloped in a faint red ring, rather irregular in shape. In addition (plate 9), one may observe a few concentrations of red light. These mounds, projecting beyond the solar disk, are called prominences. Prominences tend to hover over sunspot regions, but the sunspots themselves are on the limb of the Sun and are invisible.

The red layer is called the chromosphere. The "chromo" (from the Greek for "color") serves to remind us of its redness. The chromosphere is faint in comparison with the solar disk and normally invisible. Even during a total eclipse, it appears only for a few seconds, before it too is covered up by the Moon. Detailed observations of the chromosphere during eclipse were first made in 1869. Because the chromosphere is visible for so short a time, careful preparation is necessary if one is to observe it scientifically.

Prominences, Spicules

The chromosphere can be observed more leisurely using telescopes that artificially block the solar disk. These instruments have now been used routinely for several decades. Figure 3.4 shows one rather elaborate prominence. Its various linear and circular filamentary structures are believed to indicate strong electrical currents running within its gases (see chapter 5). Figure 3.5 shows a different type of prominence. It lasted a long time, but portions were continuously falling downward, to be replaced from above. Apparently the hot corona, invisible in this photo-

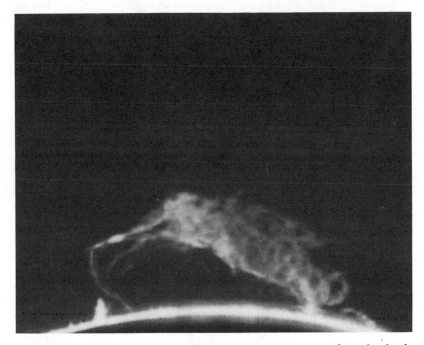

Figure 3.4. Solar prominence. This gaseous structure at the solar limb was observed at McMath-Hulbert Observatory on August 25, 1939. The Sun is artificially occulted so that only chromospheric structures are observable. This prominence rises about 100,000 kilometers above the solar surface. The gases remain at that height, despite gravity. They are supported from below by magnetic forces. (McMath-Hulbert Observatory of the University of Michigan)

graph, cools and condenses into prominence gas, but how this process occurs—and whether this is the correct explanation—is unknown.

The spicules, yet another feature, are visible when the solar disk is covered up, as shown in figure 3.6. These are flame-like structures shooting upward from the surface. Each spicule lasts a few minutes, during which time it rises some 10,000 kilometers. Then it fades from view or falls back and is replaced by other spicules. The Sun is densely covered by spicules at all times.

Figure 3.5. Prominence, December 7, 1970. Observed at Sacramento Peak Observatory, New Mexico, in the light of hydrogen, this prominence reaches heights of about 50,000 kilometers. Its shape is qualitatively different from that of figure 3.4. Individual knots of gas continuously "rain" down to the photosphere, but they are somehow replenished so that the entire prominence structure can last for hours or even days. (National Optical Astronomy Observatories)

Corona

At full eclipse, the Moon also covers up the chromosphere. As the lunar shadow deepens, the sky becomes dark, and a mysterious, rather ghostly glow appears around the Moon. It is the corona (plates 3 and 9). Its brightness is only a millionth of that of the normal Sun, so it is completely safe to view with the naked eye. In fact, it is so faint that the eye must be "dark-adjusted" to appreciate its beauty. That is, it takes time for the eye to adjust to and recognize low levels of light. At least a few minutes are needed. That is why one should not try to look at the diamond ring earlier. The eye has no time to adjust to the faint light of the corona before the eclipse is over. Dark-adjustment must be accomplished beforehand.

Figure 3.6. Spicules. These flame-like structures, observed in the light of hydrogen, rise some 10,000 kilometers above the solar surface. Spicules typically are visible for a few minutes, fade from view, and are replaced by others. This picture was taken in 1971 at Sacramento Peak Observatory, New Mexico. (National Optical Astronomy Observatories)

Figure 3.7. The corona at the eclipse of June 11, 1983, features extended "helmets." (High Altitude Observatory)

The corona is indeed a beautiful sight, well worth an extended trip for this view alone. No two eclipses are alike. Speculation about the shape and brightness of the corona abounds before each eclipse. Figures 3.7 and 3.8 show two more examples of the corona as it appears at various times. Sometimes it appears as a collection of streamers, and sometimes it features a few long "helmets" reaching far into space. The streamers and helmets grow fainter the further out one looks. Indeed, they continue far beyond our ability to observe them. The longer the eclipse and the wider the shadow path of the Moon, the darker the sky and the further out one can see the corona if the sky is clear. Also, the higher the altitude of the site and the less humid and dusty the air, the darker the sky and the further out one can see the corona. The eclipse of March 7, 1970 (figure 4.1), was unusual not only because it lasted long but also because the weather was favorable. As a result, in the United States the eclipse was visi-

Figure 3.8. The corona at the eclipse of February 16, 1980, consists of more numerous coronal loops and streamers, forming a more nearly circular corona than in 1983. The corona is photographed with a camera developed by Gordon A. Newkirk, in red light, through a radially graded filter that suppresses the bright inner corona 10,000-fold compared to the faint streamers in the outer corona. (High Altitude Observatory and Southwestern at Memphis)

ble from Louisiana to Virginia; and in Mexico, scientific expeditions saw the corona further out than anyone had anticipated.

The nonastronomer has much to see during totality. Part of the time, one wants to fix the image of the corona in one's memory: What color is the corona? How many streamers are there, and in what directions do they point? How far out are they visible? Are they uniformly bright or do they have some bright condensations near their feet? But there are many other phenomena to observe and questions to ask. Are stars or planets visible in the dark sky near the Sun? Have shadow bands traveled along the ground since just before totality? How many? How fast? These shadow bands tell us something about waves in our own atmosphere. Does the scenery appear to change? During the Alaskan eclipse of 1963, we saw some of the mountains near Mt. McKinley in sunshine, but Mt. McKinley, clearly visible five minutes earlier, was invisible in the shadow of the Moon.

All too soon, the Moon moves off the Sun. The phenomena observed before totality now take place in reverse order. The chromosphere appears briefly, the diamond ring flashes into view, and then the eye instinctively looks away from the bright Sun. The eclipse is over, at least for the people who merely wanted to see the spectacle. For the scientific expedition, this is but the beginning of the next phase of work. The data must be carefully preserved and shipped home, to be analyzed and discussed for at least a year or two.

Expedition

The members of scientific expeditions frequently do not manage to watch and enjoy the eclipse at all for one must make the most observations possible in the short time available. The time of totality is merely the central event of an effort that probably lasts at least two years.

Some decisions must be made at least a year before the eclipse. A suitable site must be selected. The decision is based on many factors: What is the chance of clouds being present? Does the center of the eclipse path fall onto the ocean? That is not good, because ships do not provide a stable platform. If the path falls on land, does it fall over a developing country? Is there a road that can handle trucks carrying food and equipment? Is an electric generator needed? Is there a source of safe drinking water? Must the equipment be protected against dust storms while being set up during the week before eclipse?

Cultural and political questions must also be taken into account. Will the inhabitants be friendly? If they consider an eclipse a bad omen, they might believe that the visiting strangers caused this bad omen. If the local population is fighting a civil war, are there insurgents who might expect to benefit by taking a hostage? The eclipse of March 1988 in the Philippines was not observed by a national U.S. expedition in part because its safety could not be ensured.

Once an expedition sets up camp, will it know accurately where it is? How well surveyed is the country? For delicate comparative measurements the observers should know their position to within some 100 meters. Will suitable satellites be available to determine positions without surveying?

The assistance of local officials is vitally important. Will the customs office guarantee that the equipment can freely enter and leave the country? A three-week embargo of scientific equipment at the customs office is a common occurrence. An unexpected six-week embargo could spell disaster for an eclipse expedition.

Two or three weeks before totality, the expedition arrives to assemble its equipment. Everything has to work during totality. There are no second chances. A detailed schedule is made for every observation, including contingencies, such as a sudden dust storm darkening the sky, a passing cloud that demands changes in the observing schedule, a corona that is brighter and more extended than normal and that demands shorter exposure times, or a malfunction in one of the cameras that must be compensated for by additional observations. The expedition must be prepared to deal with all of these things. With good planning and with everything working, the observer may finish the observing sequence a few seconds before totality ends. Now is his or her only chance to see the eclipse.

Year After

After the camp is taken down and the expedition returns home, the analysis of the data begins. Not only must the photographs be developed, but they must also be calibrated so that the brightness of all the features photographed can be measured accurately. Some of the observations during the eclipse are devoted to such calibrations, but there are always corrections to be computed and applied afterward. Did some stray light in the jerry-built equipment reach the camera? Was a part of the corona brighter than anticipated for the calibration observations? Was the chromosphere visible for less time than expected and thus were some exposure times incorrect? All these questions must be considered. Often the necessary corrections cannot be made without discussing these factors with other members of the expedition who have complementary observations.

It takes time to convert the observations into data and the data into intelligible results. Perhaps a year after the eclipse, it is possible to compare and evaluate all the observations taken during the eclipse. Results from different experiments may corroborate or challenge the initial interpretations of the data. Once this has been done, the conclusions can be refined, articles can be prepared for professional journals, and perhaps notices can be distributed to science reporters for newspapers and popular journals. Of course, the sponsor of the expedition must be suitably informed of all this progress. In the United States, this sponsor is primarily the National Science Foundation, working in collaboration with other national agencies. Still later there may be an international symposium involving participants from all the nations that mounted expeditions. Finally, the results may be presented at a session of the General Assembly of the International Astronomical Union. By this time at least two years will have elapsed and well over a million dollars will have been expended. Was it worth it? What new understanding was acquired?

Solar Atmosphere

The various phenomena associated with the eclipse begin and end quite suddenly. The bright disk disappears suddenly; the chromospheric flash disappears suddenly. The gases one sees seem to be layered. One speaks of mainly three layers of the solar atmosphere: the photosphere, the chromosphere, and the corona. Although these three regions overlap, it is convenient to consider them separate layers.

The photosphere is the "surface" of the Sun, appearing to us as the bright solar disk. It is a gaseous layer, roughly 200 kilometers thick, from which we receive most of the sunlight. Light emitted below this layer cannot escape directly; light emitted above it is relatively insignificant in amount. The temperature of the photosphere is between 5,000 and 6,000 degrees, which is too hot for solids or liquids to exist.

Above the photosphere is the chromosphere. It is red because its temperature is roughly 10,000 degrees and, seen against the dark sky, hydrogen at this temperature shines most intensely in the red, at a wavelength of 6563 Å. The same gases seen in front of the bright disk appear relatively faint. That is why Fraunhofer's spectrum of the solar disk shows a dark line at 6563 Å, that is, dark relative to the disk emission at nearby wavelengths. The "dark" 6563-Å line is not totally black but has the brightness of the chromospheric radiation.

The intensely red color of the chromosphere tells us not only that we are observing hydrogen but that the temperature is about 10,000 degrees. Hydrogen radiates as observed only if its temperature is between about 8,000 and 30,000 degrees. Therefore, "about" 10,000 degrees must be interpreted as being between 8,000 and 30,000 degrees. If necessary, information from other atomic radiations may be used to determine the temperature more precisely.

A particularly puzzling question about the chromosphere is, Why is it hotter than the underlying photosphere? We know that heat flows outward from the Sun. We also know that heat flows from hotter to cooler regions. The chromosphere, lying outside the photosphere, should be cooler. Why then is the chromosphere hotter? The answer: There must be some flow of energy other than thermal energy. Probably there are upward-traveling sound waves that turn into shock waves (equivalent to sonic booms on Earth). These then turn into heat when they reach the chromosphere.

The chromosphere is an irregularly shaped layer, as is evident from plate 9 and figures 3.4–3.6. On average, it extends about 10,000 kilometers above the photosphere, which is about the same height that the spicules reach. The chromosphere is very thin in comparison with the diameter of the Sun. Prominences frequently reach 50,000 kilometers above the photosphere. There they are surrounded by gases belonging to the corona.

Megadegrees

Most of the coronal light we see during an eclipse is ordinary sunlight scattered toward us by electrons in the corona, much as lamplight is scattered toward us by cigarette smoke. However, the corona also emits its own radiation, at selected specific wavelengths. These wavelengths do not include the pattern of ordinary hydrogen (specifically, not 6563 Å), but that fact merely tells us that the coronal temperature is different from that of the chromosphere. The observed wavelength pattern remained an enigma for seven decades, because it was not at all similar to the pattern associated with any other known atoms of the time.

The puzzle could not be solved until considerable progress was made in atomic physics, during the 1920s and 1930s. To everyone's amazement, it was found that some of the radiation comes from iron, but not from the ordinary iron atoms known on Earth. Rather, the radiation comes from iron atoms in which many electrons have been removed. We speak of highly ionized iron. Typically, between 9 and 12 electrons have been removed.

High ionization is very difficult to achieve in the laboratory on Earth. The reason lies in the structure of atoms. The nucleus of an atom is positively charged. It attracts electrons. A neutral atom has just enough electrons to balance the charge on the nucleus. For iron, the normal number of electrons is 26.

It is fairly easy to knock out one electron. Even the mild collisions among atoms in the photosphere do this easily. After a collision knocks out an electron, the resulting ion has a net positive electric charge and tends to attract a free electron. The ion moves about until it recaptures an available free electron. At any one time the photosphere contains some neutral iron atoms and some singly charged iron ions, plus at least as many free electrons. The most abundant element, hydrogen, is also partly ionized and contributes additional free electrons.

At higher temperatures, all atoms, ions, and electrons move about more rapidly, collisions among them are stronger, and more electrons can be knocked out. However, succeeding electrons are ever more tightly bound to the ions and are ever more difficult to knock out. To knock out the tenth electron of an iron atom after nine electrons are gone takes highly energetic collisions. Moreover, the highly charged ions are very efficient in attracting and recapturing free electrons. If, at any one time, many iron ions are missing 10 electrons, then there must be many energetic collisions. Such frequent, energetic collisions occur only if the temperature is sufficiently high.

The observation of coronal radiation from iron ions with 9 to 12 electrons removed implies a coronal temperature of at least a million degrees! The details of the coronal radiation were not really settled until the 1960s. Today, one considers two million degrees a typical coronal temperature. A few coronal regions may be at five million degrees, others merely at one million.

Once the high coronal temperature was recognized, however, an even greater enigma became evident: How is it possible to heat the coronal gas to a million degrees or more? The answer may involve electrical currents and short circuits in these currents, but it is still far from settled. (More on the question appears in chapter 8.)

We observe radiation from hot ions of iron. Of course, all the coronal gas is hot, not merely the iron ions. The iron is merely a minor component of the gas. It acts as a tracer for the bulk of the coronal gas, which is mostly hydrogen. The hydrogen is not observable because the hydrogen atoms have lost almost all their electrons at this temperature, and bare hydrogen nuclei do not radiate noticeably. However, the electrons removed from the hydrogen are observable. These are the electrons that scatter the sunlight toward us. They produce what we see as the corona during eclipse.

X Rays

We know from laboratory physics that gases at a million degrees or more radiate mostly X rays. Their visible radiation is merely a side effect. Is it possible that all the eclipse expeditions concerned themselves merely with a side effect of the corona? Yes, indeed! Most of the radiation from the corona is in the form of X rays.

X rays do not penetrate our atmosphere. When they impinge on a molecule or atom of the terrestrial atmosphere, they dissociate the molecule or ionize the atom. The X rays are used up in the process. This destruction of solar X rays is fortunate for life on Earth, but it is unfortunate for solar science. The X rays can be observed only from space. That is why solar science depends heavily on space observations. The enormous progress in our understanding of the solar corona was made possible by space experiments, in particular by Skylab.

More Expeditions

Why are eclipse expeditions still being organized when space observations are possible? One reason has to do with the smallest features recognizable on the Sun. Small features of the chromosphere are covered or uncovered by the Moon rather quickly. Measurement of this time interval may identify structures smaller than can be recognized by ordinary photographs. Occasionally, modest eclipse experiments can compete with some of the most expensive solar observing ventures (chapter 17).

The interesting scientific questions change with time, in part because of space experiments. A second reason for eclipse expeditions is that they may provide an important initial investigation of new questions. Moreover, eclipse expeditions can be mounted in about one year, whereas space experiments have a lead time of at least five years. Also, an entire eclipse expedition costs much less than a single space experiment. However, the relatively low cost may not be a practical advantage since the funds come from different federal agencies with highly unequal budgets: The budget of an eclipse expedition is as significant to the National Science Foundation as the budget of a space experiment is to the National Aeronautics and Space Administration. (The space agency is prohibited from supporting ground-based solar research unless it is explicitly related to a space experiment.)

One example of current eclipse experiments, one out of many, deals with the size of the Sun. In the last 10 years serious debate has developed as to whether the solar size is really constant. Might the Sun be shrinking slowly? One way to seek an answer to this question is to measure the position of the edge of the solar eclipse shadow very precisely. We already know the position of the Moon and its mountains quite precisely, and from this we can deduce the precise position of the edge of the Sun and thus the size of the solar disk. During the 1983 eclipse in Indonesia, the shadow edge was measured by several college students from the United States. They did not see the full eclipse, but they earned their travel to a fascinating country by making valuable measurements. Whether the solar size is constant is not known yet, largely because no change has been convincingly demonstrated and still more accurate measurements must be made. Ultimately, a long-lived space experiment may have to be designed to obtain an answer. However, that experiment will be expensive and must be restricted to seek answers to only very specific and well-posed questions. It takes eclipse expeditions as well as other ground-based investigations to specify these questions.

Eclipse expeditions have had one particularly famous scientific consequence. When Einstein developed his theory of general relativity, he realized that starlight passing near the Sun would be deflected by solar gravity. He predicted that stars visible near the Sun would appear to shift from their normal positions by a tiny but very specific amount. The shift would be measurable, in Einstein's lifetime, only for stars close enough to the Sun to be observable only during eclipse. Indeed, an eclipse expedition in 1919 confirmed Einstein's prediction.

Chapter 4

Active Regions

Observing Continuity

Our view of the Sun during an eclipse lasts but a moment. Even if we were to travel to every eclipse, we would observe the corona for only about three minutes a year, on average. If we observed a child for only three minutes a year, we would know so little about the child's activities that we could not possibly explain them to someone else. In the past decade we have learned that this analogy is extremely apt.

How can we observe the solar corona when no eclipse is conveniently provided by the Moon? One way is to use a coronagraph, an instrument that artificially blocks out the solar disk. Unfortunately, it cannot block out the bright sky. It is difficult to "subtract" the bright sky from photographs. Observations with ground-based coronagraphs were relied on throughout the century before space observations, but they were limited to rather brief views of the brightest, innermost parts of the corona.

X Rays and Streamers

More information can be obtained by observing the solar X rays from above the Earth's atmosphere. X rays are emitted copiously by the hot corona, but not detectably by the underlying relatively cool photosphere or chromosphere. Therefore, in X-ray images such as those in plates 2 and 4 we observe the corona and only the corona. There is no bright disk and no bright sky to confuse us as there is in visible light.

The X-ray picture in figure 4.1 shows the Sun just minutes away from the eclipse of March 7, 1970; the corona visible during that eclipse is pictured in figure 4.2. Both are pictures of the corona. How can we be sure that the two pictures correspond? They have been printed in the same orientation, so that they can be compared.

An important clue is obtained from the largest streamer in the eclipse picture, the one in the upper left. If it were extended onto the solar disk, it would hit the intense X-ray region in the upper left of the X-ray image. A second streamer, on the left of the eclipse picture, can be extended onto the disk so that it will hit the other intense X-ray emitting region, also on the left. Therefore, it appears that X-ray regions lie at the feet of coronal streamers.

Clearly, the two figures do not correspond exactly. In part, this is merely a geometry problem: The X rays are seen in front of the disk; the streamers are seen off the disk and some—for example, the isolated streamer at the bottom of the eclipse photograph—probably disappear behind the disk. A physical difference is also evident: The X rays do not appear as streamers, even if they appear near the edge of the disk. They are emitted only close to the disk. The X rays are emitted preferentially by the lowest, densest parts of the streamers. They mark only the feet of the prominent streamers.

Photosphere: Sunspots

The X rays mark regions in the low corona, several thousand kilometers above the photosphere. What lies underneath? Figure 4.3 indicates part of the answer: sunspots. The figure is a photograph of the Sun; the image is the same as the one that we can see if we project the Sun through a pinhole onto a piece of paper. Such a picture is called a "white-light" picture, as it represents all the colors radiated by the Sun. Figure 4.3 was also taken on March 7, 1970, the day of the eclipse, and oriented to match figures 4.1 and 4.2. Several groups of sunspots are apparent. The largest group, toward the left, occurs in the same position on the disk as the most intense region of the X-ray emission in figure 4.1. The position of the other two spot groups corresponds to the other two regions of intense X-ray emission. The correspondence is quite general: X-ray emitting regions always hover over groups of sunspots.

Chromosphere: Spectroheliogram

The chromosphere lies beneath the X-ray emitting regions. At eclipse it appears for even a shorter time than the corona. Fortunately, it can be observed routinely from the ground by means of a "spectroheliogram." This is a picture of the solar disk taken not in "white light" but using the radiation at only one wavelength, namely the wavelength of one of the dark lines in the Fraunhofer spectrum. A line frequently chosen is the hydrogen alpha line at 6563 Å. This radiation is so strongly absorbed that it cannot escape from the photosphere. It can reach us only from the chromosphere. Therefore, a picture using this radiation represents the chromosphere. Figure 4.4 shows a spectroheliogram taken in the hydrogen alpha line. It was taken on March 7, 1970, at the same time as the pictures in figures 4.1, 4.2, and 4.3.

The bright regions in the spectroheliogram are in the same areas as the groups of sunspots. In fact, the sunspots are easily found in the spectroheliogram once one knows where they are from the white-light picture. The same bright regions occupy nearly the same areas as the intense X-ray-emitting regions. They also occupy the same areas as the feet of the streamers seen in eclipse. Much other activity occurs in these regions, particularly flares, which occur primarily near complex sunspots.

Active Regions

Sunspots, flares, bright regions on a spectroheliogram, bright regions in X rays, the feet of coronal streamers—all of these phenomena tend to occur together. Therefore, they are known collectively as an "active region." For practical purposes, an active region is defined by the outline of the bright area on a spectroheliogram. Daily maps of active regions are routinely published. Each active region is given a number when it first appears. It may grow, change its shape, and gradually disperse in the course of weeks or months. Throughout this time, the active region rotates with the Sun and may repeatedly disappear behind the Sun. Some active regions have been well documented by solar astronomers because they have produced more flares than most other active regions and thus have become the focus of much solar research.

The term "active region" implies that its components (sunspots, flares, etc.) are causally connected. One is tempted to assume that the sunspots cause the other phenomena, but possibly both the sunspots and the other phenomena are caused by some underlying phenomenon that has not yet been properly identified. For the moment, it is sufficient to recognize the geometry: The streamers "stand" over the bright regions of the chromosphere that surround the sunspots, and the "feet" of the streamers, being the densest parts of the corona, are traced for us by X rays. All of these "hover" over and are approximately centered on sunspots in the photosphere.

Skylab

Figure 4.1 shows a photograph taken during a rocket flight. A rocket is above the atmosphere for only about three minutes and thus can provide only a few X-ray snapshots. Although figure 4.1 was certainly worth the proverbial ten thousand words, rockets cannot provide adequate information about what happens in the corona over the course of time. Routine coronal observations can only be made from a satellite that is in orbit about the Earth for days, preferably for weeks or months. Skylab, launched in 1973, revolutionized our understanding of the corona because it routinely acquired solar X-ray pictures for an entire year, the quality of which were unmatched by those obtained from rockets.

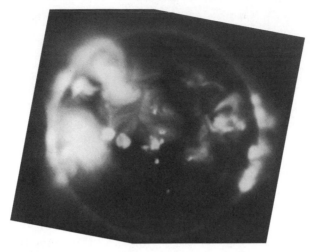

Figure 4.1. The Sun in X rays. The picture was taken by a rocket on March 7, 1970, just minutes away from a total eclipse. The Moon blocks part of the picture in the lower left. Regions of strongest X-ray emission appear almost white. (By permission, American Science and Engineering)

Figure 4.2. Corona in visible light at eclipse of March 7, 1970. The streamers on the left, when projected downward toward the disk, hit the disk where figure 4.1 shows the two most intense regions of X-ray emission. (Courtesy High Altitude Observatory)

Figure 4.3. Photosphere. White-light photo taken on March 7, 1970. The sunspots reside in the photosphere beneath the coronal regions with the most intense X-ray emission (figure 4.1). The indentation marks geographic North. The oval distortion is a camera effect. (Sacramento Peak Observatory, National Optical Astronomy Observatories)

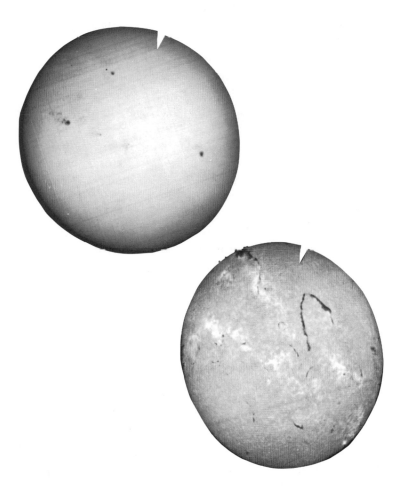

Figure 4.4. Chromosphere. This spectroheliogram was taken in the light of hydrogen, at 6563 Å, on March 7, 1970. The same spots are recognizable as in figure 4.3. Bright areas surround the major sunspot groups. They traditionally define the solar "active regions." The active regions are also regions of intense coronal X-ray emission (figure 4.1). The filament appearing dark on the disk at the top projects past the limb and appears bright against the black sky. It would be observed as a red prominence during eclipse. (Sacramento Peak Observatory, National Optical Astronomy Observatories)

67

Loops

Skylab pictures revealed not only that the corona has vastly more structure than anyone had anticipated, but also that it has a particular kind of structure. The corona consists of "loops," or arcs, of many different lengths. Plate 4 shows this structure. Notice the very long loops, connecting the main centers of X-ray emission. Each of these centers is an active region lying over a group of sunspots. Even though the spot groups are invisible in X rays, we may guess that the longest X-ray loops actually connect the sunspot groups.

There are also shorter loops, which emerge from a center of X-ray emission and return to it. If the longest loops connect sunspot groups, one may guess that the shorter loops connect sunspots within groups. Sunspots are highly magnetic. Pairs of sunspots act as if a bar magnet were placed underneath the solar surface. Indeed, the shorter coronal loops look remarkably like the patterns made by iron filings scattered on a piece of paper surrounding a bar magnet. The magnetic nature of the corona becomes apparent in this pattern of loops. Since nearly all of the X-ray emission occurs in loops, the X-ray emission of the entire lower corona is thought to be controlled by the magnetic character of the sunspot groups.

The highly structured corona demonstrated by Skylab disproved the notions of the 1960s about how the corona might be heated. At that time the heating of the corona was thought to be like the heating of the chromosphere, which is due to sound waves and shocks (sonic booms). However, shocks would produce very little coronal structure. The preponderance of coronal loops suggested that the heating is related to magnetism and electrical currents. This is a qualitatively different and substantially more complex subject than sound waves, and is still not fully understood (see chapter 8). However, the complexity also has a virtue: The new explanations account for many additional phenomena that could not be explained earlier, such as the coronal transients and coronal holes (see chapters 9 and 10, respectively).

In the course of a year's observations, Skylab transformed our understanding of the corona. Contrary to earlier suggestions that the corona is an amorphous, nearly static structure, it is actually a highly structured and dynamic phenomenon. The Solar Maximum Mission provided even further evidence of this. Before these new ideas can be consolidated, however, countless observations will have to be made over many years to come.

Temperature Structure

Skylab had yet another virtue. It revealed that the chromosphere and corona form a continuum, and permitted scientists to carry out a detailed investigation of the structure of the solar atmosphere, all the way from the photosphere into the corona.

The atmospheric structure is best described in terms of temperatures, in part because the observational methods focus on layers of specific temperatures, and in part because the important heating and cooling processes depend on temperature. Unfortunately, we cannot visualize temperatures, but we can describe them in sentences like "The temperature is x degrees at a height of y kilometers." This approach is adequate as long as one realizes that the height scale should remain flexible. It is different over sunspots, over the unspotted parts of active regions, and over quiescent regions.

The temperature of the chromosphere is estimated to be about 10,000 degrees, and to range from about 8,000 to 30,000 degrees. Intuitively, one expects the brighter areas in a spectroheliogram to be hotter than average, the darker ones cooler. Intuition turns out to be correct. However, the arguments needed to support it involve much highly detailed knowledge about atoms and their interaction with radiation. Only a few experts can actually assemble the arguments needed to show that the brighter areas in a spectroheliogram are really the hotter ones.

Spectroheligrams also show some relatively dark regions, notably the long dark "filaments." Figure 4.4 shows a very long horseshoe-shaped filament. It also shows a filament that extends beyond the limb. There, the same filament is relatively bright when compared to the dark background. If this filament were seen at eclipse, it would appear as a bright prominence. Prominences extend high into the corona. One might think that they

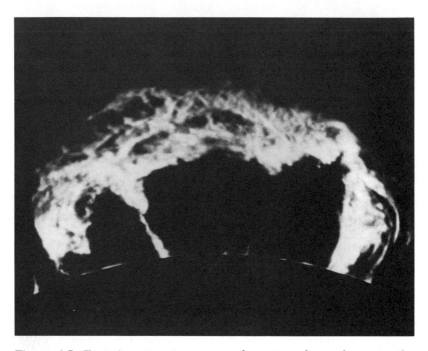

Figure 4.5. Erupting prominence caught rising above the Sun. The erupting gases have reached a height of about 300,000 kilometers. Many of the filamentary structures appear to be portions of spirals. This intricate form suggests that the gas is magnetized and has electric currents running through it, currents that are left from the time when the prominence was anchored near the chromosphere. (National Optical Astronomy Observatories)

would heat up, that the hydrogen would become ionized and unobservable. Yet their darkness in front of the disk indicates that they are among the cooler chromospheric gases. They appear to be extremely well insulated against the surrounding heat and the coronal radiation. The insulation is probably provided by the magnetization and electrical currents indicated by the shapes of many prominences (figure 4.5).

Gases hotter than about 30,000 degrees can be observed emitting at wavelengths in the ultraviolet or X-ray range. The underlying photosphere does not radiate in these wavelength ranges and causes no confusion, as the X-ray picture of figure 4.1 illustrates. Skylab and subsequently the Solar Maximum Mission observed the Sun at a very large number of wavelengths and recorded a correspondingly large range of gas temperatures.

At temperatures above about 30,000 degrees, hydrogen is ionized and cannot contribute much to radiation at specific wavelengths. It simply becomes transparent. However, helium is more difficult to ionize. When its temperature approaches 50,000 degrees, it loses one of its two electrons, but it still retains one electron. This helium ion is a strong emitter in the ultraviolet, at 304 Å (figure 4.6). Therefore, a picture of the Sun using radiation at 304 Å shows the pattern of the gas at close to 50,000 degrees. An example is the erupting prominence shown (in false color) superposed on the transient of plate 5.

Diagnostics

Strictly speaking, we can deduce from plate 5 only that helium is erupting from the Sun, and that this helium is at about 50,000 degrees. However, we know that the gas is nearly all hydrogen. We are sure that the hydrogen and helium are moving together, and that they are at the same temperature. Although the helium is a minor constitutent of the gas, it is important because it emits the radiation that we observe. The helium radiation provides information on the behavior of the entire gas. Consequently this radiation has a diagnostic function. Solar astronomers refer to it simply as a "diagnostic."

At temperatures beyond about 90,000 degrees, helium is also fully ionized and transparent. To investigate the yet hotter gases we must use radiation from a somewhat heavier element, one that can lose several electrons and still retain at least one electron to radiate. The abundance of such an element will be far smaller even than that of helium. Again, however, it must be at the same temperature as the entire gas. Its radiation serves as a diagnostic. Plate 10 represents some very simple loops observed in the light of neon. Actually, it is not enough to observe a single element because each ion serves as a diagnostic for only a relatively small temperature range, much as helium serves as a diagnostic for gases with temperatures about 50,000 degrees. A sequence of ions, from several elements, is actually used to provide diagnos-

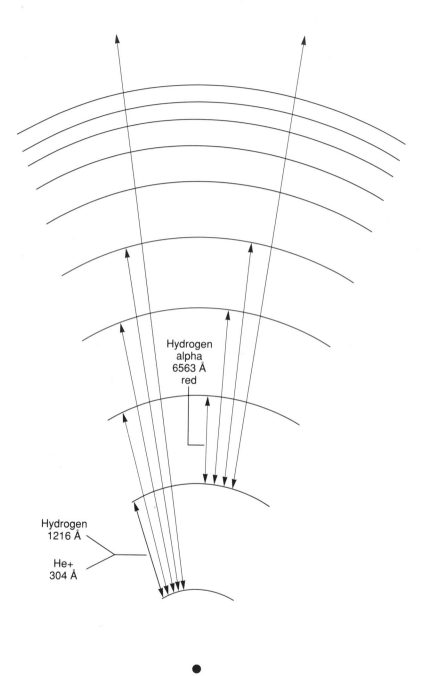

tics for ever hotter gases. The atomic structures of the relevant ions tend to be much more complex than shown in figure 4.6, but most scientists leave the complexity to the specialists and continue working with a mental picture like the one in figure 4.6.

Transition Region

Above the chromosphere is the "transition region," which is hotter than the chromosphere but cooler than the corona and roughly in the range of 100,000 to 500,000 degrees. One convenient radiator is carbon that has lost two of its electrons. It radiates at 977 Å and is characteristic of gas at 100,000 degrees. Another is oxygen that has lost five electrons. It radiates at 1032 Å and is characteristic of gas at about 300,000 degrees.

The transition region is a very thin layer. There is rather little gas there to radiate, so that the transition region is difficult to observe. However, it turns out to be an important layer for understanding the corona. No matter how the corona is heated, there is likely to be some energy that is not radiated away from the corona but is conducted as heat downward to the denser layers, those dense enough to radiate away the heat. The transition region probably constitutes this sink for the coronal heat. It must be considered a coronal "safety valve."

If the chromosphere is irregular in shape, the transition region is even more irregular. The transition region is like a thin

Figure 4.6. Atomic structure. Atoms are visualized as consisting of a nucleus with electrons orbiting about the nucleus on a select number of orbits (not drawn to scale). The neutral hydrogen atom and the He^+ (helium) ion each have one electron. Of greatest relevance for hydrogen observations is the hydrogen alpha transition in the red, at the wavelength of 6563 Å, and the transition in the ultraviolet, at 1216 Å. For He^+ it is the equivalent transition in the extreme ultraviolet, at 304 Å. Atoms emit these radiations when their electrons jump from the outer to the more central orbit. Electrons populate the outer orbit and are ready to radiate only when the gas is within specific ranges of temperature, between about 8,000 and 30,000 degrees for hydrogen 6563 Å, and roughly 50,000 degrees for He^+ 304 Å.

sheath wrapped around the outside of spicules. Obviously, it must change with time at least as fast as the spicules do, and probably faster. This makes it difficult to observe.

Atomic Notation

Some standard nomenclature must be introduced at this point. To refer to the radiations of hydrogen, helium, carbon, and oxygen described so far, we use the more concise terms HI 6563 Å, HeII 304 Å, CIII 977 Å, and OVI 1032 Å, respectively. In each case, the letter(s) stands for the element and the Roman numeral for the remaining electron that is responsible for the radiation. For example, radiation designated as HeII 304 Å is emitted by a helium ion from which one electron has been lost and in which the second electron causes the radiation (figure 4.6). Similarly, OVI 1032 Å is emitted by an oxygen ion from which five electrons have been lost and in which the sixth electron causes the radiation, with two additional electrons in the ion remaining inert.

A simple general rule to remember is this: The higher the Roman numeral, the hotter the gas in which this ion occurs. In all cases, most of the gas is hydrogen. The other elements are present in minor amounts but nevertheless provide a diagnostic by which we can learn something about the solar atmosphere.

Corona

Finally, there are diagnostic lines for the corona. Two common lines in the visible are the "red line," FeX 6374 Å, referring to gas at 1 million degrees, and the "green line," FeXIV 5303 Å, referring to 2 million degrees. These lines have led to the discovery of the high coronal temperature. But the information derived from them is limited because they represent somewhat secondary atomic processes. Ultraviolet and X-ray lines are more direct diagnostics.

One diagnostic line observed from Skylab was MgX 625 Å, which represents a temperature of 1 million degrees. Figure 4.7

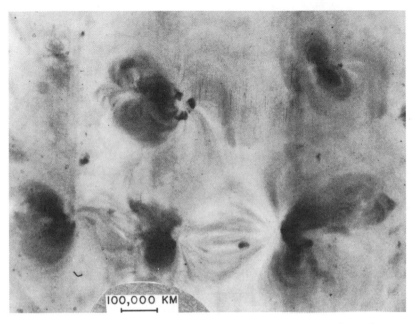

Figure 4.7. Corona observed on September 5, 1973, using the radiation at 625 Å from magnesium that has lost nine electrons. Such a picture can yield more detailed information about coronal structure than the routinely acquired X-ray picture (see plate 4). (Naval Research Laboratory and National Aeronautics and Space Administration)

shows the Sun observed in this radiation, on the day that seven active regions were visible; this was also the day that Plate 4 was recorded using a broad band of X rays. At first, the two pictures look quite similar. But there is a significant difference: The X-ray picture was taken during a routine observation, whereas the MgX picture was produced by much more select observing equipment used only on special occasions. The MgX picture can be analyzed in much greater detail and thus can provide more specific information, but it is also more difficult to obtain. Because of the similarity between the two pictures, some have argued that routine X-ray observations can be interpreted as if they were actually the more detailed observations obtained with MgX pictures, but this approach has given rise to heated debate at some scientific conferences.

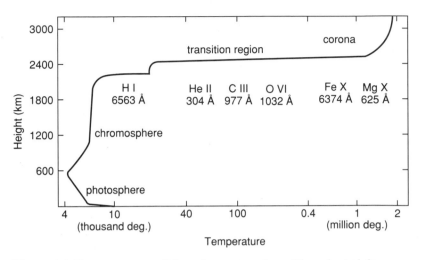

Figure 4.8. Temperatures of the solar atmosphere. The selected diagnostic radiations demonstrate higher stages of ionization (Roman numerals) at higher temperatures. Wavelengths short of 3000 Å require space observations. The heights are quite schematic because the shapes of the various regions are irregular. Much of the chromosphere consists of spicules. Much of the transition region forms a sheath around the spicules.

Figure 4.8 illustrates one way of summarizing the temperatures and related diagnostics in the solar atmosphere. There is no need to memorize the wavelengths. They indicate mainly whether the observations must be made from space (wavelengths shorter than 3000 Å) or from the ground. Figure 4.8 is based on two main principles.

First, the higher the temperature, the more destructive the collisions among atoms and ions. At the very lowest temperatures, there are no destructive collisions. For example, people occur at room temperature. People are made up of molecules, mostly water molecules. Fortunately for us, collisions at our temperature do not destroy molecules. Water can remain liquid. The Earth can remain solid. At 4,000 degrees, solids and liquids tend to evaporate and only a few molecules survive the breakup by collisions. The coolest parts of the solar atmosphere, including the sunspots, do contain a few molecules. They are gone at 6,000 degrees, the temperature of most of the photosphere. At these temperatures, many elements lose their first electron. Iron

(Fe) and calcium (Ca) are examples. At about 10,000 degrees, hydrogen becomes partly ionized. It becomes almost fully ionized above 30,000 degrees. Other elements lose more electrons as the gases become hotter. In very hot flares, one sees radiation from FeXXVI, that is, from iron ions that have lost all but one electron. Such radiation implies a temperature of about 20 million degrees.

Second, the higher the temperature of a gas, the more its various radiations can reach into the short-wavelength regime. Relatively cool objects such as people radiate only in the infrared, at very long wavelengths compared to those of visible light. Objects at a few thousand degrees, such as molten metal in a foundry or gas in the solar photosphere, radiate somewhat in the infrared, but mostly in the visible. They do not radiate in the ultraviolet or the X rays. For that reason one can photograph the Sun in the ultraviolet or X rays without being swamped by the radiation from the solar photosphere. Gases up to about 100,000 degrees radiate somewhat in the infrared and visible, but mostly in the ultraviolet. Finally, gases above a million degrees also radiate somewhat in the infrared, visible (including the coronal red and green lines), and ultraviolet, but mostly in the regime of X rays. In principle, most coronal diagnostics should involve X rays. However, it is technically difficult to isolate specific X-ray wavelengths. Therefore, Skylab provided specific diagnostics mostly in the ultraviolet. The routine pictures in X rays involved a broad band of wavelengths, which was rather difficult to analyze. Launched in 1980, the Solar Maximum Mission corrected this deficiency. Solar flares have been analyzed using diagnostics with wavelengths as short as about 2Å.

As human beings have evolved, they have become sensitive to visible light because the Sun provides it copiously and because the Earth's atmosphere transmits it. The atmosphere blocks X rays, ultraviolet, and most of the infrared. Consequently, our outlook on the universe has been quite narrow. Space observations have extended our horizons. Solar space observations have expanded them dramatically.

 PART II

THE ACTIVE SUN

Chapter 5

Sunspots and Magnetism

Symptoms

Sunspots are the most obvious symptom of solar activity. Some spots are extremely simple in shape (see figure 5.1). The darkest part of a spot is called the "umbra," from the Latin word for shadow. It is by no means as black as it looks on most photographs. It appears dark simply because most photographs are exposed for the brighter surroundings and the spot is underexposed. The umbra is surrounded by the "penumbra," a region of intermediate darkness. This region consists of radially aligned filaments (figure 5.2). Beyond the penumbra lies the ordinary photosphere. The photosphere is also structured. The most obvious feature in both figures 5.1 and 5.2 is the granulation pattern. The individual cells of the granulation change every few minutes.

When a spot is photographed routinely and frequently, many temporal phenomena become apparent. Small bright spots may appear briefly within the umbra. Waves of alternating brightness and darkness appear to run outward through the penumbra. They are largely unexplained and thus do not (yet) provide any useful information.

Simple round spots tend to lead rather uneventful lives. Scientific interest centers more on the large complex spots, such as the group in figure 3. They are the spots associated with most of the activity. In particular, the days in which spots grow and become more complex most rapidly are the days in which flares are most likely to occur. This association provides a significant clue to the cause of flares.

Unfortunately, observations of the complexity and rate of

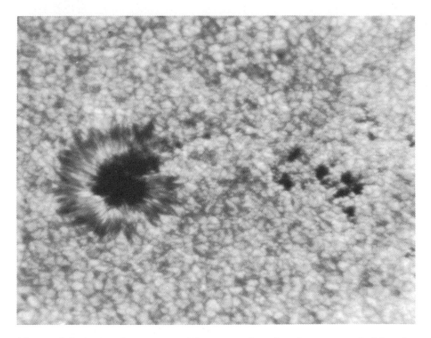

Figure 5.1. A simple sunspot. The central umbra is surrounded by the filamentary penumbra. Both are surrounded by the photospheric granulation. The diameter of the penumbra is about twice that of the Earth. It is difficult to achieve pictures with this clarity looking through the Earth's atmosphere. This photo was taken from *Spacelab 2* on the Space Shuttle *Challenger* in August 1985. The photo resolves features as small as 200 kilometers. The black pores on the right may grow into sunspots. (Alan M. Title, Lockheed Research Laboratories, and National Aeronautics and Space Administration)

change of a spot group are too inadequate to enable us to predict when flares will occur. Although spots are routinely classified in terms of their complexity, no one has been able to devise a quantitative measure of complexity that is useful for flare prediction. Frequently one ignores the complexity and reverts to thinking of spots as simple pairs of round spots.

Why Dark?

Why are the spots darker than their surroundings? The answer is that the spots are not as hot as their surroundings.

Figure 5.2. Penumbra. The highly filamentary radial structure is controlled magnetically. (Big Bear Solar Observatory)

In general, the warmer an object, the more brightly it radiates; conversely, the cooler an object, the less it radiates. Thus the relatively dark spots must be cooler than their surroundings. This argument is supported by spectra. Some molecules are identified in the spectra of spots. Molecules are totally destroyed at the ordinary photospheric temperature. They can survive at sunspot temperatures that are about a thousand degrees lower. Detailed analysis confirms this temperature difference: Typical temperatures are 4,600 degrees in the spot umbra versus about 5,600 degrees in the normal photosphere.

Convection, Granulation

Why are spots cooler than their surroundings? The answer is that they get less energy from the solar interior than does the normal photosphere.

The normal photosphere gets most of its energy from the interior by the upwelling of gases, that is, by convection. The upwelling is observable as the "granulation" occurring everywhere outside the spot in figure 5.1. The gases rising from the interior are warmer than average, and brighter. After reaching the photosphere, they radiate a part of their energy into space. On cooling, they become less bright and begin to sink again into the interior.

A movie of the photosphere reveals that the granulation is in a state of continual turbulent change. The scene has been likened to a can full of wiggling worms. A typical convection cell (seen in figure 5.1) has a diameter of several hundred kilometers, which is about the same as the distance across a large state in the United States or a country in Europe.

The solar granulation appears much less turbulent than theorists have predicted; it is almost as if the solar gases were many times more viscous than expected. Perhaps such discrepancies should not be surprising. A convection cell brings to the surface gases from a depth of several hundred kilometers and returns other gases to that depth. The gases undergo very large changes in density and temperature while they are moved about. Such motions are difficult to reproduce in the laboratory. Attempts at simulating them on the computer have demonstrated exceedingly complicated patterns of motion. They occur irregularly, appearing quiescent at times and very vigorous at others. In these motions one can also discern narrow regions of strong downdrafts, patterns that even the experts' physical intuition does not easily take into account.

The granulation pictures taken on *Spacelab 2* showed a qualitatively new phenomenon: exploding granules. Some granulation cells seem to undergo a sudden convulsion, which sends waves of changing brightness into surrounding cells. Presumably these are compression waves caused by something like an explosion in the convulsing cell. Perhaps a new upwelling caused rising gases to overshoot; when they fall back down, their impact will resemble an explosion. Whatever the explanation, up to half of all granulation cells are overrun by the waves! Surely these waves alter the granulation fundamentally, but how they do so is currently anybody's guess.

Convection is one of the fundamental physical phenomena not

yet understood. Solar convection dominates many other uncertainties in understanding the Sun. It is also important for the science of aerodynamics because it portrays a regime not easily studied on Earth. Therefore, much effort is put into photographing the motions involved in solar granulation and tracing their changes. Unfortunately, the normal motions of our atmosphere blur the photographs of granulation quite significantly. Photographs from space are a marked improvement. *Spacelab 2* produced 6,400 solar pictures showing entire sequences with unprecedented clarity. But the observations lasted only two days, which is too brief a period to evaluate the pictures, discover the exploding granules, and follow up on this discovery. Alternatively, electronic means may correct ground-based telescopes for atmospheric blurring. Such a procedure is still in the experimental stages and needs to be developed for several more years before it can be used for routine observations. Even that may not solve the problem of granulation, because we shall still be getting only a superficial picture of the underlying convection. Much theoretical extrapolation to greater depths will still be necessary.

Why Less Hot?

Why are spots cooler than their surroundings? Convection provides heat to all of the photosphere except sunspots. Since there is no convection in spots to provide energy efficiently the surface of the spot cools off. At lower temperatures, it radiates less. When the temperature has decreased sufficiently, the radiation loss is just balanced by the available heat supply (by radiation and conduction). That is where the temperature remains. The details of this adjustment have not yet been fully evaluated, in part because the necessary computations must be made on supercomputers that have only recently become available.

Magnetism

Why is convection inhibited in sunspots? The answer is that the sunspots are strongly magnetized, and magnetism inhibits convection. Spots are dark because they are cooler; they are cooler because convection is inhibited; and convection is inhibited because of the spots' magnetic properties. Of course, that again begs the question: Why is the sunspot magnetic? Briefly, the answer is we do not know. All attempts to trace the magnetism to its origin have been based on convection, and the conclusions are still speculative. For the time being, then, solar scientists are tackling problems that do not depend on this information.

The magnetic properties of the solar gases are regarded as the cause of all solar activity. Sunspots are the most visible and flares the most dramatic symptoms.

Zeeman Splitting

How do we know that the solar gases are magnetized? In the laboratory, the wavelength radiated by atoms changes slightly when the atoms are near a strong magnet. Radiation normally at one wavelength is split into radiation at two or more slightly different wavelengths. This is called Zeeman splitting, named after Pieter Zeeman, who discovered the effect in 1896.

In the solar spectrum, magnetism splits some single Fraunhofer absorption lines into two. Fraunhofer observed the spectrum of light taken from the entire Sun and could not have recognized the phenomenon. To observe Zeeman splitting easily, one must first isolate the light from a sunspot. Figures 5.3–5.4 and plate 11 show how this is done. One produces an image of the Sun and only allows some of the light to pass through a slit. That light is then dispersed and its spectrum displayed. The result is the spectrum of that part of the solar image that happens to fall onto the slit. If the image of a sunspot falls upon a portion of the slit, the result is the spectrum of a sunspot. Its Zeeman splitting can be seen in plate 11.

Plate 11 merely shows one magnetic effect on the solar spectrum, namely the splitting of a single Fraunhofer line. In addition, the radiations received at the two shifted wavelengths turn out to be polarized in opposite senses. The sense of polarization informs us whether the spot is a north or a south magnetic pole.

Figure 5.3. The McMath Solar Telescope on Kitt Peak, Arizona. Sunlight is gathered by a 2-meter (80-inch) sun-tracking mirror atop the tower. The light travels along the polar-axis tunnel, which extends far underground, and is finally imaged near ground level (see figure 5.4). (National Optical Astronomy Observatories)

Most important, the splitting and the polarization together constitute proof that a sunspot is highly magnetized.

Visualization

The magnetization of solar gases poses a mathematically difficult and usually insoluble problem. Therefore, even the experts try to visualize what happens and present their work in terms of pictures, using computations as a backup. Such visualization turns out to be the easiest way to achieve some familiarity with magnetization.

The most useful picture is obtained from a bar magnet and iron filings. Look at the left side of figure 5.5. Imagine iron filings on a table surrounding a bar magnet. The magnet aligns the filings into an easily recognizable pattern.

Figure 5.4. Solar image. The image can be moved so that any desired part of the image falls onto a slit, whence the transmitted light passes through a spectrograph. When a sunspot is directed onto the slit, plate 11 results. (National Optical Astronomy Observatories)

A bar magnet happens to be a convenient object. However, an electromagnet provides a better analogy for the Sun. An electromagnet is really nothing but a coil of wire with electrical leads to bring and remove electrical current. Each end of the coil corresponds to one of the poles of a bar magnet. The iron filings

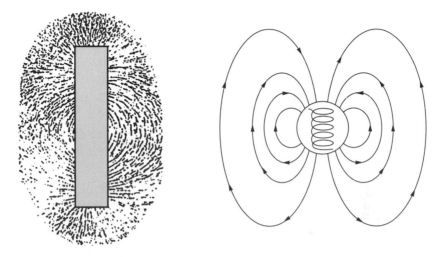

Figure 5.5. Magnetic patterns. On the left, iron filings are aligned by a bar magnet. On the right, "field lines" replace the pattern of iron filings, and a schematic electromagnet, a coil of wire, replaces the bar magnet. The circle might represent the Earth. The field is said to be stronger where the field lines are closer together.

line up around a coil with sufficient current just as they do around a bar magnet.

The electromagnet lines up the iron filings no matter what the coil is made of. All that is needed is a current that runs in a coil-like pattern. The current might run through copper wire or through any other substance—even a gas—that is capable of conducting electrical currents. This is the situation in fusion machines. It is also the situation in the Sun. The hot solar gases constitute very good electrical conductors. Electrical currents are plentiful on the Sun.

Magnetic Fields

The iron filings respond to a magnetic force. To visualize this force, one speaks of a "magnetic field." The "field" is created by electrical currents, but it can be felt where there is no current. For instance, the iron filings respond to the electromagnet even when the magnet is some distance away. The magnet need not touch the iron filings to make them move. We speak of the magnet "creating a magnetic field" everywhere outside the magnet. Each iron filing is said to "feel" the magnetic field at its own position. It reacts to this field by lining up with the field.

When all the iron filings line up, each according to the magnetic field at its location, then the ensemble of aligned iron filings displays the pattern of the magnetic field. This pattern helps us to visualize the magnetic field. If one connects all of a set of iron filings that are aligned end to end, one obtains a smooth curve. That curve is called a "magnetic field line," or simply a field line. For each set of iron filings aligned end to end, one obtains a different field line. The right side of figure 5.5 shows a few field lines surrounding an electromagnet.

A close analogy to an electromagnet is provided by the magnetic field of the Earth. The Earth's field is largely created by electrical currents deep within the Earth. It is this field that aligns the needle of a compass at the Earth's surface. One can assign a direction to the magnetic field, as shown by the arrows attached to the field lines in figure 5.5. Each arrow indicates the direction in which a compass needle would point if it were located at the position of the arrow.

Sunspots, prominences, and coronal structures provide information about solar magnetic fields in much the same way that the alignments of compass needles on Earth provide information about the terrestial field. These solar features allow us to visualize the solar magnetic fields because their gas patterns resemble magnetic patterns. The gas patterns can be photographed and are easily visualized. Figure 5.6 provides two examples. The shapes of these loops indicate an electromagnet hidden underneath the feet of the loops. When solar rotation turns them into view, the electromagnet turns out to be a pair of sunspots. McDonald's Arches (plate 10) also hovered over sunspots.

A magnetic field is not only oriented in a particular direction; it also has strength. The stronger the field, the more effectively a compass needle is lined up with the field. A drawing of magnetic field lines informs us where the magnetic field is strongest, that is, where the field lines are relatively close together. In figure 5.5 the field lines are closest together near the magnetic poles. In-

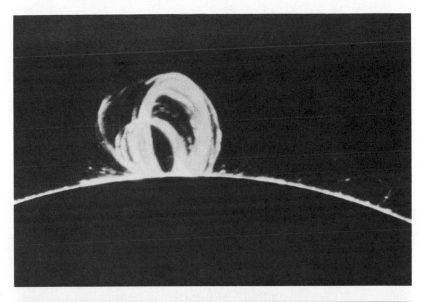

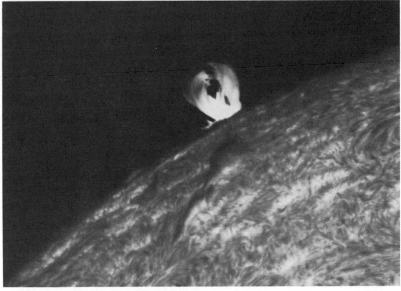

Figure 5.6. Two simple prominences, observed in the light of hydrogen. The prominences consist of gaseous loops resembling the shape of magnetic field lines near an electromagnet. Such field lines are expected near a pair of sunspots. In these edge-on views of the Sun, the sunspots are hidden. The solar disk is blocked in the first picture. The spectroheliogram of the disk in the second picture shows the great complexity of the solar atmosphere, including innumerable spicules and several dark filaments. (National Optical Astronomy Observatories and Big Bear Solar Observatory)

deed, the Earth's field is strongest near the magnetic poles. Zeeman splitting of the sunspot spectrum (plate 11) is used to measure the magnetic field strength in sunspots.

A convenient unit to use in measuring the strength of a magnetic field is the gauss. The Earth's magnetic field is about 0.5 gauss at the equator. The field measured near a strong bar magnet may be a thousand gauss or more, but it weakens rapidly with increasing distance. Megagauss fields are created in some laboratory conditions. A typical magnetic field of a sunspot measures about 1,500 gauss. The strength is not spectacular compared to that of a bar magnet, but the volume occupied by the spot field is. The field has about the same strength throughout the umbra, that is, over a volume larger than the volume of the Earth. Therefore, it is more appropriately compared with the Earth than with a bar magnet. The magnetization of a sunspot is over a thousand times greater than that of the Earth. The electrical currents creating the magnetic field must also be enormously stronger than that within the Earth. In fact, the currents in a sunspot amount to several trillion amperes!

Sunspot Model

It is frequently said that a pair of sunspots acts as if a bar magnet or a horseshoe magnet were placed under the solar surface. A rather good sunspot model can be made by varying this theme only slightly. Imagine that the electromagnet in figure 5.5 is really a coil of wire. Bend it as shown in figure 5.7. Arrange it so that each end of the coil faces upward just below the photosphere. Bend the field lines also so that each field line emerges near one end of the coil and reenters at the other end. Only the field lines should extend above the solar surface. Replace the coil by conducting gases. Replace the pair of coil ends by a pair of sunspots.

Figure 5.6 shows that this simple model of a sunspot is remarkably good. These prominences surely delineate a spot magnetic field. The sunspot model of figure 5.7 also suggests that field lines leaving the edges of sunspots are directed almost horizontally. These field lines are believed to run along the radial

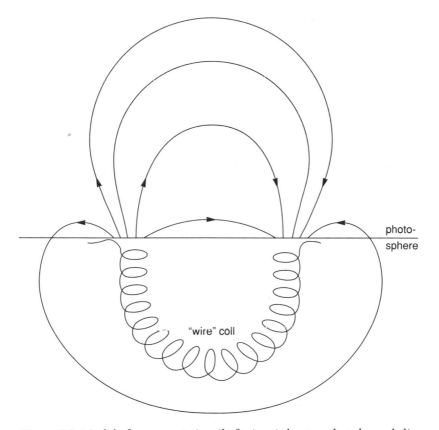

Figure 5.7. Model of a sunspot. A coil of wires is bent so that the ends lie just below the photosphere. The ends constitute a pair of sunspots. The corresponding magnetic field lines are shown. In reality we see only the portions above the photosphere and the wire coil is replaced by electrically conducting gases.

filaments that constitute the penumbra (figure 5.2). The penumbra provides us with one more method of visualizing the magnetic field.

The "recipe" for constructing a sunspot pair involves one major uncertainty: We do not know how the electrical currents travel through the solar interior. Some currents may run directly from one spot to the other, as implied by the model. But some currents may diffuse from one spot toward the interior while other diffuse currents find their way to the second spot. Without knowledge about the interior electrical currents, we cannot "explain" solar magnetization and solar activity at greater length.

Foreshortening

Figure 5.8 shows a rather simple sunspot as it disappears from view during solar rotation. When the spot is near the limb (edge of the disk), only the limbward part of the penumbra is in full view. The nearer side is highly foreshortened. Evidently the sunspot resides in a hole. The sunspot photosphere lies a few hundred kilometers below the general photosphere. The penumbra constitutes a sloping surface connecting the two photospheres. Its slope is consistent with the slope of the fields expected from the sunspot model of figure 5.7.

Sunspots turned toward the limb during solar rotation gradually disappear from view. Conversely, sunspots brought into view during solar rotation become recognizable only a day or so after limb passage. The change in group complexity that may occur behind the Sun is unpredictable. One is anxious to observe that complexity on reappearance because it suggests how much solar activity will be associated with the spot group. One must wait for the group to rotate well past the limb to identify the new structure.

Sunspot numbers and sizes are monitored regularly. The "Zurich" sunspot numbers are frequently cited. These numbers involve corrections for sunspot foreshortening and for the number of spot groups. Such corrections make the Zurich sunspot numbers subject to some controversy. Many geophysicists consider solar radio emission a less ambiguous measure of sunspot activity.

Coronal Loops

The magnetic fields visualized in figure 5.6 are observed by the hydrogen alpha emission of chromospheric gases. Similarly, coronal magnetic fields can be visualized by the emission of coronal gases in soft X-rays. Once again, look at plate 4 showing

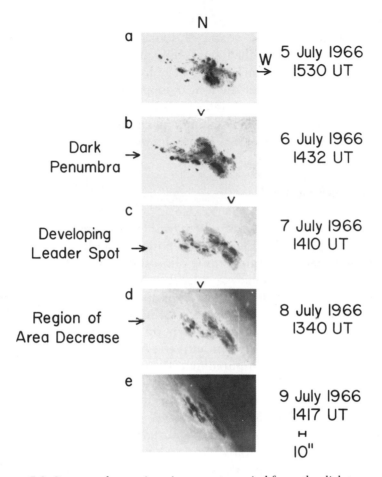

Figure 5.8. Sunspot depression. A sunspot carried from the disk toward the limb becomes foreshortened near the limb. The sunspot photosphere lies below the normal photosphere by several hundred kilometers. The penumbra constitutes a sloping surface connecting the two photospheres and is observed more easily on the limbward side of the spot. The spot group evolves during the four days of observation. The pairs of arrows point out some of the changes. The foreshortening of the leader spot on the last day precludes a good measure of its size. (Courtesy P. S. McIntosh, Space Environment Laboratory, National Oceanic and Atmospheric Administration)

seven active regions or figure 4.7. These observations show hot coronal gases in the shape of loops. Their general pattern strongly implies that the loops follow magnetic field lines.

Some of the shorter loops emerge from and reenter the same active region. Many of the longer loops extend from one active region to another. These long loops suggest that the entire corona is magnetic, although only parts of the corona contain enough hot gas to be observable as loops.

Hairy-Ball Model

There is a way to check whether the loops really follow the magnetic field lines. Figure 5.9 is a magnetogram, a map of the magnetic fields emerging at the photosphere from the five active regions shown in figure 4.7. Clearly, each active region is bipolar: Many of the fields emerging on one side of an active region return on the other. But not all. Computers have been used to visualize all of the coronal magnetic field, but especially the field that connects different active regions. The result is shown in figure 5.10. The computed field lines look remarkably like the loops in figure 4.7!

A portrayal of the magnetic corona as in figure 5.10 is referred to as a "hairy-ball" model. In general, hairy-ball models resemble the observed coronal loops remarkably well.

Hairy-ball models have two serious limitations. First, they are based on only the larger magnetic patterns observed on the photosphere. They ignore small-scale structures such as the shapes of spots within an active region. Second, they are computed on the assumption that there are no electrical currents in the corona. Since flares and all other coronal activity require coronal electrical currents, the hairy-ball model can give no information about coronal activity.

Electrical Currents

Electrical currents on the Sun can be recognized only indirectly, by a twisting of the prominences and loops that visualize the magnetic field. Figures 4.5 and 5.11 show examples of such

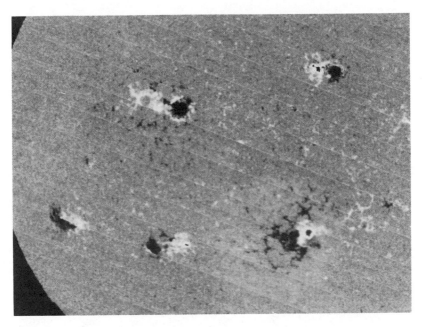

Figure 5.9. Magnetogram. This compilation of Zeeman measurements all over the solar disk on September 5, 1973, shows the photospheric magnetic fields emerging from and returning to five active regions. The coronal connections between these regions appear in figure 5.10. (Courtesy W. Livingston, National Optical Astronomy Observatories)

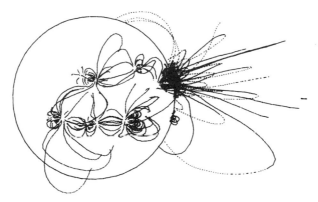

Figure 5.10. Hairy-ball model. The magnetic pattern of the corona deduced from figure 5.9 by computer closely resembles the coronal loops observed on September 5, 1973, which are shown in plate 4 and figure 4.7. (Courtesy Randolph H. Levine)

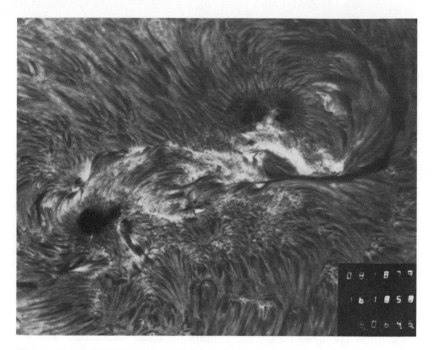

Figure 5.11. Active region of August 18, 1979. The swirling patterns almost certainly betray electrical currents of billions of amperes. (Big Bear Solar Observatory and *Solar Physics*)

twisted structures. It would be very difficult to make a model of the magnetic fields and compute the electrical currents, but those swirling patterns most certainly reveal substantial electrical currents.

The currents running in the chromosphere and corona may be quite large. A complex active region such as in figure 5.11 forces billions of amperes to run up into the corona and return downward, partly in the same active region, partly somewhere else. When new spots emerge within a spot group, the coronal magnetic pattern must change. Electrical currents amounting to billions of amperes must be added to or removed from the corona. Imagine trying to short-circuit a current of several billion amperes. If you succeeded, you would surely create quite a spark! Might such a short-circuit in the corona be observed as a solar flare (see chapter 11)? What would cause the short-circuit (see chapter 12)?

⊙ Chapter 6

The Solar Cycles

Butterfly Diagram

The best-known manifestation of the solar cycle is the changing number of sunspots visible at any one time. This number fluctuates from month to month, but on average it reaches a maximum about every 11 years. There are several standardized ways to measure the number of spots. The most widely used is called the Zurich sunspot number. It takes into account the sizes of spots, their grouping, and their reduced visibility near the solar limb (figure 5.8). The published Zurich sunspot number is usually a monthly average. It ranges from 0 at a very low sunspot minimum to about 200 during an intense maximum.

The position of newly appearing spots also changes during the sunspot cycle. Shortly after sunspot minimum, new spots appear rather far from the solar equator, typically at latitudes 40 to 50 degrees north and south. Once created, each spot or group stays at that latitude until it disappears. Subsequent spots appear somewhat nearer to the equator. They also stay at the latitude where they were formed. This trend continues for several years. About 11 years after sunspot minimum, new spots appear quite close to the solar equator, typically at latitudes about 10 degrees north and south. This is the time of a new sunspot minimum. During some minima, there are days in which no spots can be observed. When new spots appear, they appear again at latitudes 40 to 50 degrees, and start the next cycle. During other minima, spots of the new cycle appear before the last spots of the previous cycle have run their course.

The butterfly diagram (figure 6.1) portrays the changes over

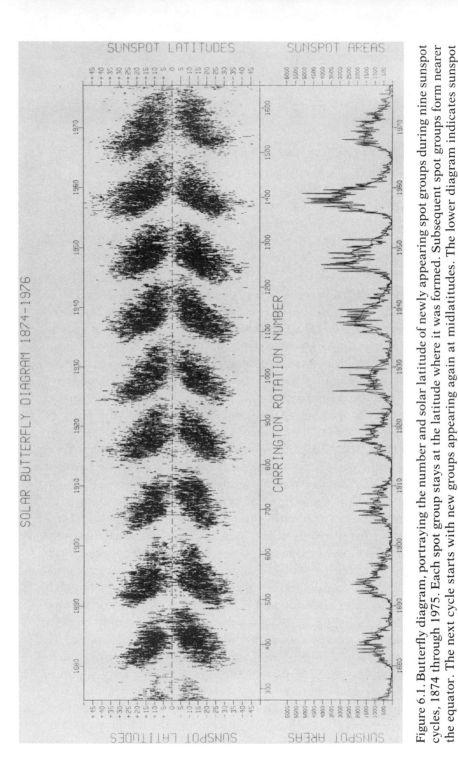

Figure 6.1. Butterfly diagram, portraying the number and solar latitude of newly appearing spot groups during nine sunspot cycles, 1874 through 1975. Each spot group stays at the latitude where it was formed. Subsequent spot groups form nearer the equator. The next cycle starts with new groups appearing again at midlatitudes. The lower diagram indicates sunspot areas averaged monthly and seasonally. (Courtesy Royal Greenwich Observatory)

time of both the number and the position of the spots. Every dot in the diagram shows the latitude and time of appearance of a spot group. All the groups appearing in one 11-year cycle constitute one "butterfly." The most recent complete cycles lasted from 1965 to 1977 and from 1977 until the end of 1986. Neither of these cycles lasted exactly 11 years.

What Next?

The 11-year cycles are not particularly regular. Most obvious in the butterfly diagram is the irregular "length of the wings." The maximum of 1958 reached particularly high latitudes. The duration of the individual cycles is also irregular. More often they last about 10 or 12 years rather than the average 11 years.

The maximum number or area of spots to develop at the various maxima is also irregular. The lower part of figure 6.1 shows the area of sunspots with time during the years 1874 through 1975. The curves are very irregular, so that one may arrive at different conclusions using the monthly or the seasonal averages. Nevertheless, the maximum of 1958 was obviously a very large one.

The graph in figure 6.1 ends with 1975 because the Solar Department of Greenwich Observatory was closed soon thereafter. This was the situation when plans were laid for the Solar Maximum Mission, to be launched during the maximum of 1980. How intense would that maximum be? What clues did the data up to 1975 provide? One possibility, as indicated by figure 6.1, is that the sunspot maximum grows for several cycles, then starts over again at a low value. If so, 1980 would be much like 1969. Was it?

Figure 6.2 shows the sunspot numbers from 1610 to 1985. The graph is smoothed even more than figure 16.1, but clearly the maximum of 1980 was very intense, the most intense in three centuries except for 1958.

The next sunspot maximum is expected in the period 1990 to 1991. Can one now predict how high that maximum will be? Many people try to find longer cycles in the data and try to use these cycles to predict the maximum height. For example, if one

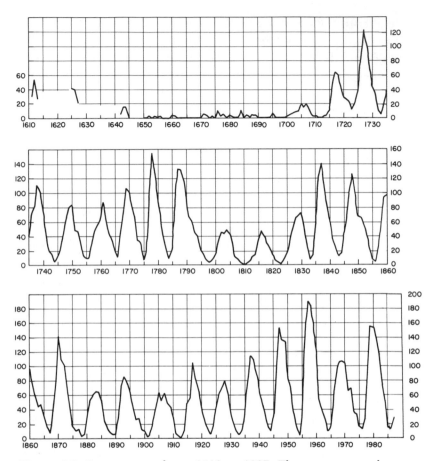

Figure 6.2. Sunspot numbers, 1610 to 1987. The sunspot numbers at maximum change from cycle to cycle. Some viewers discern a saw-tooth pattern in the maxima. Activity increases sharply after the minimum of 1986, suggesting a strong maximum in 1991. The period with essentially no spots, 1645 to 1715, is named after E. Walter Maunder, who tried to publicize this phenomenon in 1894. (Courtesy John A. Eddy, University Corporation for Atmospheric Research)

looks at the variation of the peaks of cycles in figure 6.2, optimists may find a cycle of about 86 years in duration. This cycle has a "saw-tooth" shape: The maximum sunspot number increases gradually from one 11-year cycle to the next until it plummets at the end of an 86-year cycle, only to start over again.

Of course, one has to consider that this cycle length is also irregular. If one accepts this type of analysis, two possibilities can be entertained. If 1969 was anomalously low and 1980 was the peak of a cycle, then the maximum will now plummet, so that the next maximum will be a relatively low one. If 1969 began a new cycle of growing maxima, then the maximum of 1991 might surpass even that of 1958, and we should really worry about the maximum of the year 2001.

The amplitude of the next sunspot maximum is significant for humans because the activity associated with sunspots affects communications, electric power supplies, the welfare of numerous Earth-orbiting military and civilian satellites, and much more (see chapter 14). Realistically, it is virtually impossible to predict the next cycle. The irregularities in the shapes and amplitudes of the peaks in figure 6.2 are greater than any regularities that may be hidden in the data. As of 1988, the spot numbers for the new solar cycle have increased as rapidly with time as they did in the intense previous cycle. The summer of 1988 even featured a naked-eye sunspot! It would be wise to anticipate a very high maximum for 1989–91 and frequent terrestrial disturbances as a result.

Skylab

A dramatic example of our inability to predict the sunspot cycle was provided by Skylab. Skylab was abandoned in space when the third crew of astronauts departed in 1974. Even then there were plans to refurbish Skylab once the Space Shuttle became operational. However, it was uncertain whether Skylab would still be in orbit at that time, because of atmospheric friction. In time, friction with the highest parts of our atmosphere tends to lower the orbit of any satellite. The friction increases rapidly as the orbit becomes lower. Finally, a fairly sudden "reentry" into the atmosphere destroys the satellite. Skylab's orbit was initially high enough so that friction was by and large unimportant. However, if there was an unusually high number of sunspots, the associated active corona would emit enough x-radiation to heat the Earth's atmosphere and to raise it to the height of Skylab, causing friction and the loss of Skylab.

Could one predict the number of spots for the next few years, until the shuttle could raise the orbit of Skylab to a safe altitude? The long delay in the first shuttle flight made such a venture impossible before the 1979–81 maximum. A high solar maximum would certainly lead to Skylab destruction before 1981.

It so happened that the sunspot minimum of 1977 was unusually low. No spots appeared for many months. Thus some were optimistic that the next spot maximum might also be low, sufficiently low for Skylab to survive and to be refurbished at leisure. This was wishful thinking at best.

When the next cycle started, it started at a rate beyond all expectations. In fact, the maximum reached in December 1979 was the second largest during the past three centuries. The upper reaches of our atmosphere promptly rose in response to the X-rays from the active corona. The friction on Skylab became severe. By early July 1979, it was clear that Skylab would soon reenter the atmosphere. Because Skylab was so large, pieces were expected to survive until impact on the ground. Where would impact occur? At best, reentry could be predicted to within about two hours. In that time, Skylab circled the Earth. It might reenter at any part of its orbit. Pieces might fall anywhere beneath its orbit. The public was reassured by the fact that the previous Skylab orbits had been over ocean more than half the time. But there were some anxious minutes around the world.

Skylab reentered the atmosphere on July 11, 1979, after almost 35 thousand orbits around the Earth. Pieces weighing up to 850 kilograms dropped onto Australia, dispersed over a path 6,400 kilometers long. No one was injured. The demise of Skylab was a direct consequence of solar activity.

Hubble Space Telescope

Astronomers have pinned much of their planned research on the Hubble Space Telescope, to be launched by the Space Shuttle. It is a billion-dollar venture that was to be launched during 1986. The *Challenger* disaster has postponed launch until February or March 1990, near the time of sunspot maximum. As of August 1988, NASA expects to squeeze out of the Space Shuttle a launch altitude of 611 kilometers, rather than the normally cited capability of 593 kilometers. At that altitude the telescope should survive even an intense maximum. But in case the maximum is an extreme one, no booster is available. Perhaps launch will have to be postponed yet another few years? The cost of maintaining the telescope on the ground is some 100 million dollars a year. More important, the equipment on the telescope, designed in the early 1980s, is rapidly becoming obsolete. The telescope is designed to be refurbished in space about once a decade. Will it need to be refurbished on the ground before launch?

Twenty-two-Year Cycle

An increasingly important property of the sunspot cycle is related to the magnetic character of a spot pair.

Most pairs of spots are aligned in an east-west direction. In most pairs, the spot that first comes into view during solar rotation tends to be slightly larger. Its magnetic field dominates a larger area around the spot. Therefore, it is useful to distinguish the "leading" spot and the "following" spot in a pair, where these terms refer to the direction of solar rotation.

If one observes all the spot pairs visible at one time, one finds that all the leading spots in the northern hemisphere have the same polarity, and all the leading spots in the southern hemisphere have the opposite polarity.

During the next 11-year cycle, all the orientations are reversed. If the leading spots in the northern hemisphere are north poles during one 11-year cycle, then they are south poles in the next one; they become north poles again in the cycle after that, and so on. In fact, one identifies a new spot cycle not only by the appearance of new spot pairs at higher latitudes but also by their reversed polarity. The spot pairs in the southern hemisphere are nearly always reversed relative to the northern pairs (see figure 6.3). An exception occurs when a new cycle starts in

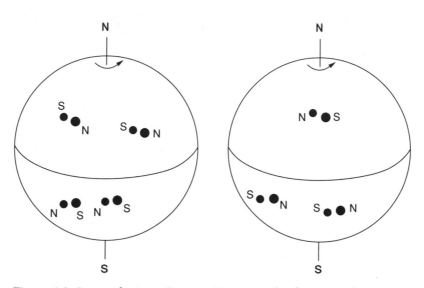

Figure 6.3. Spot polarities. For one 11-year cycle, the spot polarities are organized as on the left, for the next cycle as on the right. The magnetic pattern repeats with a cycle of 22 years.

one hemisphere a few months before it starts in the other hemisphere, which is one more manifestation of the irregularity in the solar cycle.

The pattern of magnetic orientations recurs about every 22 years. The solar cycle is really a 22-year magnetic cycle. The 22-year cycle can be detected in several other, less obvious phenomena, including photospheric velocities and various structural patterns in the chromosphere and corona. Just as the zones with sunspots migrate from midlatitudes to the equator in 11 years, these additional phenomena migrate from the poles to the equator in 22 years. The spots merely join the 22-year migration during its second half.

The first of these 22-year phenomena to be discovered, about a decade ago, is a large-scale velocity pattern (see figure 6.4). Any one segment of the photosphere rotates slightly too fast for about 5 years, slightly too slow for the next 6 years, and so on. The pattern of these changes in velocity exhibits a 22-year cycle. "Slightly" too slow or fast means just that: The changes in rotational velocity are merely 3 meters per second, twice the

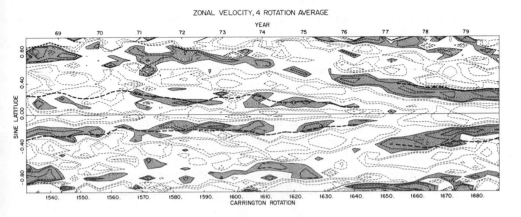

Figure 6.4. Twenty-two-year velocity cycle observed at Mount Wilson. Solar rotation changes very slightly with time and latitude. At each latitude, rotation is slightly faster than average for a few years, shown as dark in the figure, and then slightly slower than average, with an 11-year cycle, here from 1969 to 1979. At any one time, there are two zones of slightly faster rotation, two slightly slower, between pole and equator. This pattern drifts from the pole to the equator in about 22 years. During the second half of the 22 years, zones with sunspots and active regions join the migrating pattern. They occur at the latitudes where rotation changes from slightly faster to slightly slower than average. The velocity pattern and the sunspots appear to be merely different manifestations of the 22-year solar cycle. (Courtesy R. Howard and *Astrophysical Journal*)

normal walking speed on Earth but less than 1 percent of the velocity of solar rotation. They could be discovered only when spectrographic instruments could be maintained stable and reliable to this degree for years at a time, a significant technical achievement.

Superficial Blemishes?

The solar cycles present an enormous challenge to scientists. Although the motions of figure 6.4 are observed only on the solar surface, their large scale suggests that they originate deep in the solar interior. The sunspots are more local phenomena, but the associated electrical currents may also reach deep into the solar interior (see chapter 7).

The solar cycles are probably only surface manifestations of some complex phenomena in the interior. One speaks of a "solar dynamo" operating in the solar interior. However, this impressive name is merely a coverup for our substantial ignorance. The problem is that we must deduce the interior phenomena from secondary effects that are observable at the solar surface. Compared to the dynamo that drives the solar cycle, the sunspots are mere surface phenomena. Galileo and his contemporaries were quite right in using the word "blemish" to describe them.

Magnetogram

The magnetic cycle must involve more than merely sunspots, because other parts of the photosphere are also magnetic. As noted in chapter 5, the magnetic fields emerging at the photosphere can be mapped. Full-disk "magnetograms" have been constructed for several decades. These maps of the magnetic fields are based on measurements of Zeeman splitting over the entire solar disk. Magnetic fields as weak as 10 gauss can be recorded. A magnetogram is shown in figure 6.5. It was taken on March 7, 1970, the day of the solar eclipse, and corresponds to the spectroheliogram in figure 4.4 and to the X-ray image in figure 4.1. Clearly, magnetic influence extends over a substantial portion of the solar disk.

The magnetic fields recorded in routine magnetograms are largely confined to active regions. When there are no active regions, on some days near solar minimum, the magnetogram may be essentially blank. However, more detailed data show all solar gases magnetized to some degree.

Emerging Magnetic Flux

A convenient measure of solar magnetization is "magnetic flux," which is simply the product of field strength times the photospheric area penetrated by the field. It is basically what the eye

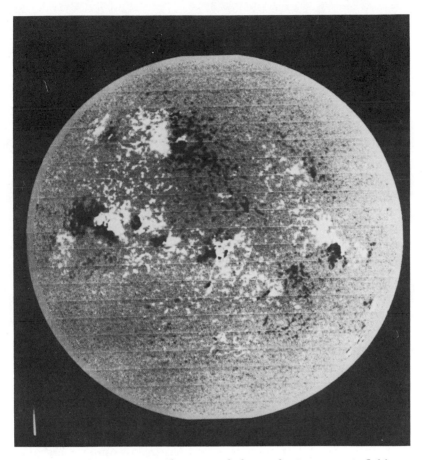

Figure 6.5. Magnetogram. This map of photospheric magnetic fields on March 7, 1970, is based on measuring the Zeeman splitting at many parts of the disk. White and black indicate opposite magnetic polarities. The positions and extent of the magnetized regions are similar to those of the active regions in the spectroheliogram in figure 4.4 and of the X-ray-emitting regions in figure 4.1. The active regions are bipolar in the same sense as sunspots, in that they reverse across the equator: Northern active regions are divided into white on the left, black on the right; southern ones are reversed. The polarity pattern is even more pronounced in the magnetogram of September 5, 1973, in figure 5.9. Average magnetic field strengths recorded in active regions range from 10 to 100 gauss. The instrument does not record sunspots. (National Optical Astronomy Observatories)

sees on a magnetogram, independently of magnetic polarity. Since every field line leaving the photosphere must reenter it somewhere, each field line contributes twice to the magnetic flux.

Observers developed the notion of magnetic fields "emerging" from the photosphere, because the appearance of a new pair of sunspots looks like two objects rising from the interior and becoming visible upon passing through the photosphere. "Emergence" implies a decidedly one-way phenomenon. Most magnetic fields emerge relatively quickly, within hours or at most a few days, and have clearly recognizable polarities. Then they disperse relatively slowly. If dispersal is sufficient, oppositely polarized magnetic fields are cancelled and magnetic flux decreases. So far, magnetic flux has only occasionally been observed to sink back down through the photosphere. However, some observers now argue that much more magnetic flux actually emerges and then promptly sinks back down, so that it is easily missed. If so, this is one more factor pointing to the restlessness of the Sun.

Sunspots grow as long as new magnetic flux emerges near them. Spots not continually rejuvenated disperse gradually. This dispersal of photospheric magnetic fields has its counterpart in the corona. The coronal loops observed in soft X rays tend to be sharply outlined when they first appear. They become increasingly indistinct as they age, giving at least the appearance of gradual dispersal.

Emerging magnetic flux is a useful way to measure the solar cycle. It is somewhat analogous to the sunspot number but also takes into account the sizes of sunspots and the active regions. Near sunspot maximum, when large spots and large active regions are born frequently, the emerging flux is large. Near sunspot minimum, it is small. However, emerging magnetic flux is not a sufficient measure of the solar cycle. During the 1970s, two unexpected additional phenomena were discovered: intense but very localized magnetic fields in the photosphere (see chapter 7), and X-ray bright points.

X-ray Bright Points

The Skylab images in soft X rays show "points" of X-ray emission. Several dozen X-ray bright points may occur simultaneously on the Sun. They can be recognized on every Skylab X-ray image reproduced in this book. They appear all over the Sun, even well away from active regions. They are relatively short-lived, lasting from a fraction of an hour to several hours. If a movie is made of the X-ray images of the Sun, with one frame taken every 90 minutes, which is the Skylab orbital period, the X-ray bright points make the Sun appear to sparkle.

The X-ray bright points brought attention to another new phenomenon. In the photosphere beneath every X-ray bright point resides a small but intense bipolar magnetic region. It is similar to a small and close pair of sunspots, except that these tiny spots are not cool and dark. On the contrary, the X-ray emission suggests that the gases immediately above the photosphere are unusually hot.

The magnetic flux associated with any one X-ray bright point is not large. However, there are many bright points, and they appear frequently. On average, they constitute almost as much emerging flux as do sunspots and active regions. Interestingly, the bright points apparently account for more emerging flux during sunspot minimum than during maximum. If this is true, the magnetic solar cycle must be viewed in a qualitatively new light—not merely as a cycle in the emerging flux, but also as a cycle with respect to the size of the magnetic units that emerge, with relatively more small units emerging during sunspot minimum than during maximum. Most theorists consider this possibility disturbing but have not really grappled with it yet. Owing to the demise of Skylab and the lack of frequent X-ray images in the interval between Skylab and the Solar Maximum Mission, X-ray bright points have not been properly monitored for one entire solar cycle.

The observational evidence bearing on the solar cycles is still coming in. Meanwhile, it is comforting to know that other stars also have spots, coronae, and flares. Some stars appear to experience spot cycles rather like the Sun does (more on this in chapter 15). Will stellar data help to explain the solar cycles? A "wait and see" attitude is appropriate.

Chapter 7

Dynamo

Flux tubes

As already mentioned, the 1970s produced two observational developments bearing on the solar cycle: the X-ray bright points and the structure of the photospheric magnetic fields. Routine magnetograms such as those in figures 5.9 and 6.5 are for magnetic fields averaged over several thousand kilometers on the Sun. About a decade ago, J. Harvey and other solar observers concluded that this average is very misleading. When he examined the spectrum of radiation from the tiniest fraction of the solar photosphere that he could isolate, he found a spectrum composed of two parts. Most of the light of this spectrum comes from gases with very weak magnetic fields. But roughly 1 percent of the light comes from highly magnetized gases, with fields of at least a few hundred gauss.

The highly magnetized gases constitute what are called magnetic flux tubes, or simply flux tubes. The fields are directed vertically. The field strengths are difficult to measure but are usually considered to be as strong as those of sunspots, say, 1,500 gauss. The tube diameters are usually estimated to be about 100 kilometers. These flux tubes are not dark, like sunspots, because they are thin enough to be heated efficiently from the sides. On the contrary, when observed with sufficient detail the flux tubes may look brighter than the surrounding photosphere: One looks not only down into the flux tubes but also at their walls, which are below the normal photosphere and, therefore, hotter and brighter than the photosphere.

The flux tubes provide essentially all the magnetic flux emerging from active regions. Sunspots may begin as aggregates of flux tubes. Flux tubes occupy roughly 1 percent of the area of an active region. Many flux tubes occur beyond active regions. They seem to control the motions of spicules (figure 3.6). Some flux tubes occur under X-ray bright points. Clearly, flux tubes are to be reckoned with.

The locations of the flux tubes are related to the "supergranulation," which is a convection pattern on a larger scale than the ordinary granulation. These convective motions appear to sweep the flux tubes toward the vertices of the supergranulation. The supergranulation cells also reach deeper into the Sun than do the ordinary granulation cells.

Figure 7.1 shows a sketch of what the magnetic configurations in an active region are imagined to be like. The magnetic geometry above the photosphere is quite uncertain. The effect of the photospheric flux tubes is still recognizable in the radiations from the chromosphere, but not in those from the corona. Apparently the flux tubes fan out at some intermediate height.

Flux tubes are difficult to detect and even more difficult to study in any detail. A telescope with resolution 10 times better than currently available will make an enormous difference in our understanding of the flux tubes and the solar cycle. A space-based telescope with this capability has been under discussion for some years (see chapters 17). Observations from the ground can also be improved. A single flux tube has actually been detected and shown to move about using an electronically corrected mirror installed at the National Solar Observatory on Sacramento Peak, Sunspot, New Mexico. Computer correction of solar images is also promising. But these techniques are still in an experimental stage and are not yet suitable for a detailed study of flux tubes.

Magnetic Forces

Thin flux tubes and sunspots may look different, but theoretically they have a very important property in common. Since the field strengths and the generally vertical direction of the fields are similar in flux tubes and sunspots, so are the magnetic

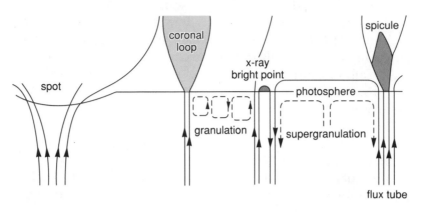

Figure 7.1. Magnetic structures. Supergranulation cells are larger than granulation cells (figure 4.1), reach deeper, and sweep magnetic flux tubes to their boundaries. X-ray bright points occur over tiny emerging magnetic flux tubes. Spicules (figure 3.6) may shoot up from other flux tubes. Field lines reaching the corona may be filled with gas and become observable as X-ray loops. The photosphere is depressed in the sunspot.

forces. Consequently, flux tubes and sunspots behave similarly. Together, they constitute most of the surface manifestations of the magnetic solar cycle. Thus it is fortunate that they can be dealt with in much the same way in any attempt to explain that cycle.

Magnetic forces are demonstrated in many science courses. Typically, the experiment involves running an electrical current through a wire while the wire is in the magnetic field of some magnet (see figure 7.2). When the current begins to run along the wire, the wire moves. The motion demonstrates that a force is acting on the wire. The force is caused by the simultaneous presence of a current running along the wire and a magnetic field directed across the wire.

The same magnetic forces act on the electrically conducting gases of the Sun. When solar electrical currents run across a magnetic field, they exert a force on the gas. Unless there is some other balancing force, the magnetic force moves the gas. Coronal transients, the sudden blowing-off of a part of the co-

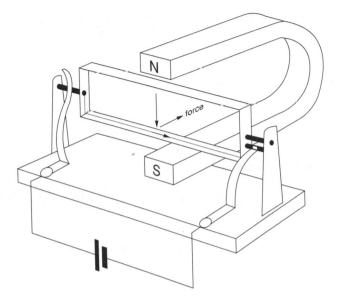

Figure 7.2. Magnetic forces in the laboratory. Here, the magnetic field is vertical between the poles of a bar magnet. A current runs along a wire on the lower half of the rectangle, which can rotate on its axis. A force is exerted on the wire at right angles to both the magnetic field and the current. The rectangle will start to turn on its axis. The circuit is fixed so that current runs only when the wire is below the axis. This forms a very simple direct-current motor.

rona shown in plate 5 and figure 9.2, are caused by such magnetic forces.

The model of a sunspot (Figure 5.7) involves electrical currents encircling the main magnetic flux tube; that is, the currents run across the magnetic field much as they do in the laboratory experiment. Some force must be acting on the current-carrying gases just as on the wire in the experiment. This force is directed outward and can be imagined to act like a pressure. The "magnetic pressure" is high inside the spot, relatively low outside. The high magnetic pressure within the sunspot helps it to withstand the gas pressures acting from outside. Consequently, the spot is adjusted so that the interior gas pressure is less than that outside. In turn, the gas density is also less inside than outside. The lower gas density has a clear observational consequence: One can see somewhat deeper into a sunspot than into the normal photosphere. Therefore, one sees a sunspot as a depression in the photosphere, as in Figure 5.8. The same is true of any photospheric flux tube, but the depression within narrow flux tubes allows one to see the hotter walls of the flux tube.

Magnetic Buoyancy

The decreased density within sunspots and flux tubes has another important consequence. The decreased density is an adjustment that preserves pressure balance, but it creates another imbalance, for example, among forces in the deeper layers of the sunspot model (see figure 5.7), where the magnetic field lines are more nearly horizontal. There, the lower gas density inside the flux tube makes the flux tube act rather like a helium-filled balloon in the Earth's atmosphere: A deficiency of gravitational force allows it to rise.

A helium-filled balloon rises rapidly. A freely floating flux tube would also rise rapidly. It would rise through the interior in a matter of hours and pop out of the photosphere in the course of a few minutes. In fact, the observed emergence of magnetic flux takes hours and sunspots remain for days. Something is wrong. Why are there any magnetic fields left inside the Sun? And why does a sunspot pair stay where it is rather than promptly floating up and away? The answer appears to be that the sunspot is anchored somewhere down below.

The magnetic anchoring must be quite reliable. The anchor apparently releases flux tubes only gradually, in the course of days, weeks, or even months. Perhaps this is fortunate. What would happen if the anchoring were to come loose suddenly? Might that cause enormous, destructive solar flares? Young stars are very much more active than the Sun (see chapter 15). Perhaps some sudden magnetic releases can occur in young stars. Perhaps they did occur when the Sun was young (see chapter 15), but such magnetic relief became unnecessary in the current, older, more staid Sun.

Frozen-in Field Lines

"Floating flux tubes" and "magnetic anchors" seem to give the impression that magnetic fields are physical objects that one can hold and work with. Yet, magnetic field lines are defined as imaginary patterns that summarize for us how iron filings might orient themselves. Such a mental leap must be justified. Indeed, this justification has become a cornerstone of much of astrophysics. It is the basis for the discipline called magneto-hydrodynamics or, less frighteningly, MHD. Its "father" is generally considered to be the Swedish physicist Hannes Alfvén.

The justification starts with a terrestial experience: A wire carrying current to a household appliance may heat up. If the current is forced to run through a very thin wire, the current heats the wire enough to sap the energy intended to reach the appliance and, usually more important, the heat may melt the wire. The remedy is to use a thicker wire. The more dispersed the current, the less heat is produced.

Very large currents flow on the Sun. Flux tubes may carry billions or even trillions of amperes. But the currents flow in patterns (gaseous "wires") with enormous cross sections, so that the current per unit cross section is minute. For instance, a current of a trillion amperes flowing through a sunspot carries only roughly a millionth of an ampere through each square centimeter cross section. No significant heat is generated by such relatively weak current densities.

Figure 7.3 shows how one can visualize the behavior of magnetic fields when no heat is generated. Field lines that behave as if stuck in the gas are transformed from an abstract concept into a property of the gas.

One speaks of field lines as being "frozen into the gas." This description is valid as long as the currents run through large enough areas for heat production to be negligible. In most astronomical situations, currents indeed run through such large areas that frozen-in magnetic fields are a very good approximation.

Whenever currents generate no heat, then the magnetic field lines associated with the currents take on a physical role, such as magnetic buoyancy or magnetic anchoring. Figure 7.3 dem-

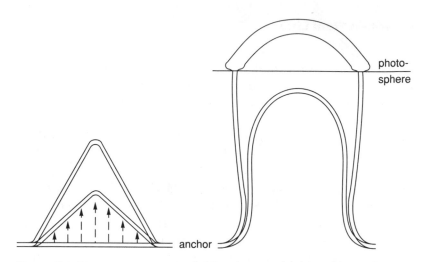

Figure 7.3. Frozen-in magnetic fields. This is a (highly schematic) vertical section of the outer part of the Sun. On the left, vertical gas motion, shown by arrows, distorts some initially horizontal and straight magnetic field lines. The figure shows two "snapshots" of the field lines at subsequent times. The magnetic field lines act as if they were stuck onto the gas. They rise with the gas as if they were "frozen in." On the right, the field lines are shown as they might look at two still later times, when subsequent gas motions have carried them yet further upward. The last snapshot shows the field lines when they emerge from the photosphere, possibly forming a pair of sunspots and a coronal flux tube. Once the field lines are nearly vertical, gas within them may drain downward. The emerging magnetic flux brings to the surface only very little gas.

onstrates both, at least with a little imagination. Indeed, imagination is essential because phenomena like emerging magnetic flux can be simulated mathematically only after extreme simplification.

An explanation of the solar cycle must include the origin of the magnetic flux tubes, the preferred magnetic orientations of the spots, and the reversal of the magnetic orientations every 11 years. These three requirements are listed in order of increasing theoretical difficulty. Even the easiest of them is still poorly understood.

Magnetic Anchors

Where are most flux tubes anchored? What mechanism provides new flux tubes to replace those that are dispersed?

The outer third of the Sun is convective (plate 7). The turbulent convective motions help to carry heat outward. Flux tubes are strongly buoyant almost throughout this volume and there is no apparent way to anchor them there. However, near the bottom of the convection zone, the buoyancy is rather weak and even a fairly weak anchoring force would suffice. Eugene Parker, of the University of Chicago, suggests that flux tubes there may inhibit convection and cause a "thermal shadow." The gases above the flux tubes become slightly cool and dense. The extra overlying gas presses on the flux tubes and can keep them from rising despite their buoyancy. Therefore, the lower part of the convection zone would make a good magnetic anchor.

Deeper in the Sun, energy is transported entirely by radiation. In this "radiative zone," flux tubes may float neutrally or even tend to sink. This region would also make a good anchor. But a further problem arises there.

If flux tubes emerge and disperse, how are "new" flux tubes created? The left side of figure 7.3 suggests how. Ideally, motions are needed that can remove one strand from a large sheet of flux tubes. Where might one find suitable motions to separate single flux tubes from adjacent ones? They certainly exist in the convective zone, but also just below the convection zone. There some of the strong downdrafts created in some convection cells may continue to travel downward, reaching thousands of kilometers into the radiative zone. These highly directive motions may be very effective in isolating some flux tubes from neighboring ones.

In view of the requirements of a good anchor together with suitable gas motions, it is surmised that the flux tubes are anchored near the radiative-convective boundary, roughly 0.2 million kilometers below the photosphere.

But another problem arises now: Newly created flux tubes must rise through the entire convection zone before emerging from the photosphere. The right part of figure 7.3 is really over-simplified. While flux tubes rise, they suffer the buffeting and distortions due to dozens of convection cells. When they emerge as a spot pair, one might expect their magnetic orientations to be random. Yet that is contrary to what is observed.

The magnetic alignments of spot groups and active regions suggest control by large-scale field lines at the depth of the magnetic anchors. Strands are removed from this main magnetic field and somehow maintain their orientation as they rise until they emerge as sunspots or portions of active regions.

Magnetic Reversal

Eleven years later, the polarities of the spot groups are reversed. Presumably the large-scale interior field lines have also been reversed. How is that possible? One thing is clear: The Sun has not been turned around. The large-scale field has not been turned around with the Sun. Instead, the field topology must have changed locally. Individual pieces of the large-scale field lines must be turned around separately. This demands a breakdown of the frozen-in approximation. Electrical currents must be dissipated by being forced to run through sufficiently small areas. How?

No one has a good theory for the dissipation of solar electrical currents. But current dissipation apparently happens frequently in several astrophysical situations. On the Sun, current dissipation appears to cause coronal heating, transients, and flares (see later chapters). Therefore, theoreticians tend to shrug their shoulders and simply assume that current dissipation happens in a manner convenient to their own theory.

Several theories for the solar cycle and the 11-year magnetic reversal have been formulated. Some depend mainly on sketches of magnetic fields; some involve very elaborate mathematical models. None are easily visualized on paper because three dimensions are needed to do so. However, all the theories agree that two familiar phenomena are necessary to help turn the magnetic field lines around: rotation and differential rotation.

On Earth, cyclones and hurricanes consist of twisting motions in the atmosphere. The clockwise or anticlockwise direction of the twist is determined in part by the Earth's rotation. These winds would not occur if the Earth were not rotating. Precisely such winds are needed on the Sun in order to twist and turn around the solar magnetic fields. Therefore, the solar rotation is considered an essential ingredient of any explanation of the solar cycle. In fact, the planets provide a confirmation of that expectation: Several planets have retained magnetic fields for billions of years, even though the fields are continually being dissipated. All these planets rotate fairly rapidly. Their rotation appears to be necessary for the continual regeneration of the magnetic field.

Differential Rotation

The twists imposed by the solar equivalents of tornadoes and hurricanes turn the solar magnetic field lines into any and every direction. They cannot provide the observed order in the sunspot polarities. That is the role of differential rotation.

The Sun rotates slightly more rapidly at the equator than near its poles. Gases at the equator make about 10 rotations while gases at midlatitudes rotate 9 times and gases near the poles rotate only about 7 times. This implies a very large-scale shear in the gas motions. It creates large-scale orientation in the magnetic field—just what is needed.

Figure 7.4 shows what happens to an initially irregularly shaped field line after it has participated in differential rotation. The parts of the field line at the equator have been carried forward the furthest. The parts north and south of the equator lag behind. They lag far enough behind so that most of the field line is oriented in an almost east–west direction. If some such field lines, well below the surface, rise buoyantly and without further twisting, as in figure 7.3, they yield spot pairs and active regions lined up in an east–west direction, with opposite polarities in the two hemispheres, as observed. Clearly, solar differential rotation provides the essential ingredients needed to explain the order in the solar magnetic fields and in the solar cycle.

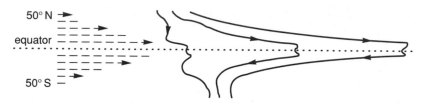

Figure 7.4. Effect of differential rotation. In this surface view of the Sun, an initially horizontal but otherwise irregularly shaped field line is "frozen into" gas with a horizontal flow (rotation) that is fastest at the solar equator. Two later snapshots show how large portions of the field line tend to align in an east–west direction. The aligned field has opposite directions in the two hemispheres. If such aligned fields occur well below the surface and portions rise as in figure 7.3, then this accounts for the ordered polarities of sunspot pairs (figure 6.3). The field lines are closest together where the field is most nearly aligned. That means the field is strongest there. It is the most buoyant. Thus the most aligned portions of the field lines are also the portions most likely to rise and form aligned spot pairs.

Solar Dynamo

The solar dynamo has been defined as the deep-seated phenomenon that causes the surface manifestations of the solar cycles. The ingredients of this phenomenon, rotation and differential rotation, are permanent features of the Sun. They suggest that a "machine" in the solar interior reverses itself after about 11 years. In fact, the name "dynamo" derives from some comparable machines on Earth.

Current theories for the dynamo do not naturally predict a magnetic reversal after 11 years. A judgment is needed as to which theories are attractive and worthy of further pursuit. To be considered reasonable, a theory must at least permit an 11-year cycle and a butterfly diagram, even if it cannot predict them. Even more attractive is a theory that can also allow the solar cycles to vary gradually over the decades and occasionally to turn off, as in the Maunder minimum. In fact, some theories are currently pursued precisely because they permit the "intermittency" of the Maunder minimum. The notion of intermittency, in contrast to predictable behavior, is currently in vogue

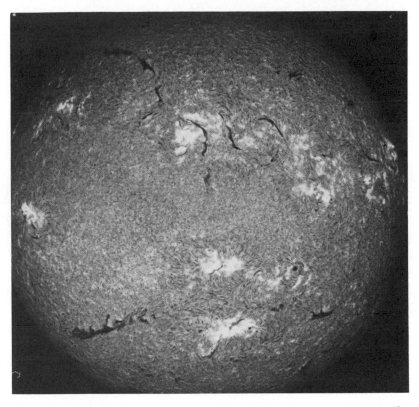

Figure 7.5. Whole-disk spectroheliogram, taken on June 4, 1968, in the light of hydrogen alpha by Patrick McIntosh using a 5-inch refractor. The continual and unpredictable evolution of solar surface structure greatly complicates the recognition of an interior dynamo machine. (Courtesy P. S. McIntosh, Space Environment Laboratory, National Oceanographic and Atmospheric Administration)

and is applied to many problems of physics, including convection. It may well apply to solar cycles also.

Possibly no present theory is on the right track. Figure 7.5 shows what the Sun really looks like. How can we extrapolate from a complicated and continually evolving exterior to a "simple" interior machine? More important, much fundamental knowledge is lacking. We do not understand convection. We cannot specify how magnetic field lines can be disconnected so that motions can turn them around. We are only beginning to put together a believable explanation of the solar differential rotation. The solar dynamo may remain unexplained for many years.

This rather negative outlook should not overshadow the accomplishments to date. Everyone now agrees that magnetic fields are important for the Sun, for space physics, and for astrophysics. The subject of magnetohydrodynamics is well established. It has not long been this way. Hannes Alfvén, the scientist who first recognized the enormous importance of cosmic magnetic fields, encountered much resistance from the astrophysical community, which did not wish to accept his ideas: Alfvén tells of a paper he wrote in the 1930s predicting a galactic magnetic field of only a billionth of a gauss. Today we routinely accept galactic field values at least a thousand times stronger. Yet at that time Alfvén could not publish the paper because his estimate of the magnetic field was considered implausibly large. There has been considerable progress since then.

Chapter 8

Skylab, Coronal Loops, and *Einstein*

Sonic Booms

The high temperature of the corona was recognized in the 1930s. It was immediately clear that radiation from the relatively cool photosphere cannot heat the corona. Physicists have known for centuries that heat will not flow from a cool area to a hot one of its own accord. This observation has been embodied in some basic physical concepts about the amount of order and disorder (entropy) in the Universe. Accordingly, the photosphere cannot heat the corona by radiation. Somehow, energy other than radiation must reach the corona from the photosphere.

One of the important sciences of the 1940s was hydrodynamics. A topic of interest, because of its practical application, was high-speed airflow past airplanes and, for very fast planes, the formation of sonic booms, more properly called shock waves. This interest led to the development of the first widely accepted theory of coronal heating, presented by L. Biermann in the 1940s. It was based on high-speed gas flows and shock waves that could be expected to emerge from the photospheric granulation (figure 5.1) and the underlying convection.

Shock Waves

Solar convection is a rather irregular process wherein masses of rising or sinking gas occasionally bump into each other. These collisions create considerable noise, which is much like thunder on Earth, but has an extremely low frequency, roughly one cycle per minute. For comparison, a tone of "middle C" has 400 cycles per second. We would not hear the solar sound waves but would feel them as gusts of wind.

The solar sound waves can travel over a long distance. In particular, they can travel upward out of the photosphere and into the chromosphere, where the gas density is lower. Sound waves traveling into a region of lower gas density become stronger. The wave motion becomes more violent. When the sound waves have ascended far enough, their crests catch up to the wave troughs and the waves turn into shock waves.

On the Earth we hear shock waves passing us as sonic booms. The booms rattle windows because they bring a sudden increase in air pressure. The booms also "rattle" all the atoms; that is, they heat all the material they pass over. The heating by terrestrial sonic booms is too rare to be significant. But the frequent heating by shock waves on the Sun might be significant. Biermann proposed that most of the shock waves traveling upward from the photosphere heat the chromosphere. A small fraction of the energy would travel higher and heat the corona.

The idea of coronal heating by shock waves was widely accepted for 30 years! Few theories live that long. However, the 1970s brought a whole array of opposing arguments. For instance, it was said that the shocks must involve gas motions, but that no such motions have been detected in the corona. More important, the coronal structure observed on Skylab X-ray pictures could not be explained by shock waves.

Skylab Again: Magnetic Structure

Skylab X-ray images showed a corona shaped by magnetic fields. This qualitative conclusion has led to the quantitative "prediction," outlined in the next few paragraphs, that the solar corona should have a temperature of about 2 million degrees, as observed. In fact, any magnetically controlled stellar corona should have about this temperature. Remarkably little information about the Sun is needed. Most of the information derives from well-known classical physics.

The "prediction" of the coronal temperature was made long

after the temperature had been deduced from observations. In hindsight, one might have expected the magnetic structure, given the temperature. However, pre-Skylab theory focused on different problems. As so often happens in science, especially for astronomical phenomena, new observations demonstrated much more structure and significant detail than we had anticipated.

Orbiting Solar Observatory

One important clue to the heating of the corona was provided by a satellite called OSO-7. OSO stands for Orbiting Solar Observatory, a series of satellites with equipment designed specifically to observe the Sun. The first such satellite was launched in 1962. OSO-7 was launched in 1971 and operated for two years. It could provide pictures of a coronal loop taken in the light of one of the coronal spectral lines. Unfortunately, it did not demonstrate the pervasiveness of coronal loops.

OSO-7 pictures showed a coronal loop to be hottest at its top. Since heat always flows toward cooler regions, one must deduce that other energy, heating the gas, is somehow provided at the top (see figure 8.1). Shocks might provide this energy. However, heating at the top is contrary to expectations for shock heating. Instead, one can imagine several magnetically controlled processes that preferentially heat the tops of loops (see chapter 9). Fortunately, all give nearly the same result for deriving the loop temperature. So little information is needed that our ignorance of the heating mechanism does not matter, given merely the dominant role of magnetic fields.

Loop Structure

The magnetic structure of the corona has an important consequence. Gas is magnetically confined. It can flow easily only along a magnetic field line, not across it. Any lateral distortion of the field lines, of the sort shown in figures 7.3 and 7.4, would require flows or gas pressures that are not generally available in the corona.

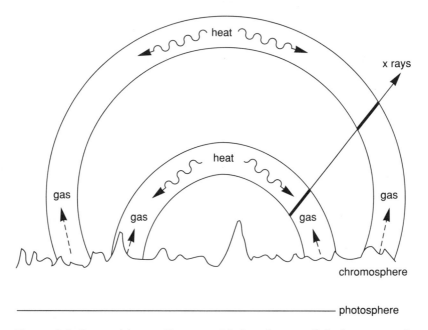

Figure 8.1. Coronal loops. Heat provided at the top of the loops travels downward. If the heat is not radiated away, it reaches the chromosphere and makes additional gas evaporate into the loop. The larger loops typically reach 50,000 kilometers above the photosphere. Temperatures might be 2.0 and 2.5 million degrees at the tops of the larger and smaller loops, respectively, and about 0.5 million degrees less at the feet. Each loop contributes to the observed X rays according to its thickness, gas density, and temperature.

The "frozen-in approximation" allows us to think of field lines as being attached to the gas. For the magnetized corona, it is more convenient to consider all gas as being attached to some coronal magnetic flux tubes, essentially coronal loops, as in figure 8.1. The magnetic field lines provide the "framework" that holds up the gas of the loop.

Where does the gas in a loop come from? Since gas can flow only along the magnetic field lines, all gas on one flux tube must have entered the corona at its feet, where the flux tube dips into the chromosphere. When gas enters at the feet, other gas on the same flux tube moves in response until the gas density is about the same all along the flux tube, albeit slightly less near the top.

The more gas that enters the feet of a flux tube, the higher its density throughout the flux tube.

The gas supply at the feet of any flux tube is practically inexhaustible. Since magnetism pervades all of the corona, one might think that all of the corona would be filled with gas. However, observations show otherwise. Only flux tubes observed as X-ray loops are well filled with gas. The emission of X rays is proportional to the square of the gas density. Therefore, a flux tube with a gas density three times larger than its neighbor is nine times more easily observable. A modest contrast in gas density causes a large contrast in the X-ray image. The space between observable X-ray loops is not empty, but merely several times less densely filled with gas, so will be unobservable on routine Skylab X-ray images.

The intensity of the observed X rays provides an estimate of the gas density in a coronal loop. A typical density obtained in this way is a few billion hydrogen ions (and an equal number of free electrons) per cubic centimeter. This is only about a millionth of the density in the solar photosphere. That this ratio is small comes as no surprise, because the brightness of the corona visible at eclipse is also merely a millionth of the photosphere.

Temperature Control

The magnetic nature of the coronal gas has a second essential consequence: Heat also flows only along magnetic flux tubes, not across them. Heat is carried by free electrons, and electrons are constrained to move along the field lines as is all of the gas.

Suppose now that energy is somehow delivered at and is heating the top of a loop, as suggested by the observations of OSO-7. Where can it go? Radiation is not efficient because radiation is proportional to the square of the gas density and is lowest where the density is lowest, at the top of the loop. Instead the heat is conducted downward, along the loop. Sufficiently far down, the density is great enough to radiate heat away as fast as it arrives. A steady state is set up, with energy continually being added at the top, flowing downward along the loop, and being radiated away nearer the feet. Since heat can only flow from hotter to cooler gases, the top of a loop must be maintained slightly hotter than the feet. The requirement fits the observations of OSO-7. In fact, the temperature difference is small enough so that one can still think of a single temperature characterizing an entire loop.

We now ask: At what loop temperature can radiation cool the feet in the same time that heat conduction from the top heats the feet? The radiation takes a few hours, depending on the density and temperature of the gas. The conduction also takes a few hours, depending on the same two factors and on the length of the loop: It takes longer to flow down a long loop. If one assumes that the two times are equal and if one knows the loop length and the gas density, then one deduces a unique value for the temperature.

The "predicted" value of the temperature is roughly 2 million degrees. "Roughly" means a range of 1 to 5 million degrees, depending to a small extent on the chosen gas density and length of the loop. The small range allows us to speak of "the" coronal temperature as being about 2 million degrees.

The attractively simple prediction of the coronal temperature requires only very basic knowledge, namely the knowledge that heat flows along magnetic flux tubes and the knowledge of how atoms radiate and electrons conduct heat, and the demonstration by Skylab that the corona is structured magnetically. The remarkable generality of this result was first recognized in a publication by R. Rosner, W. Tucker, and G. Vaiana, all then at Harvard University.

Heat Conduction

Terrestrial experience helps to show why the length of a coronal loop influences its temperature. Daily experience suggests that heat conduction is efficient when regions of high and low temperature are adjacent to each other, but is rather slow when heat needs to be conducted over large distances. At a beach on a sunny day, the energy of the Sun may heat the top layer of sand so that it hurts the soles of our bare feet. Fortunately, only a little shuffling of our feet reveals that the sand further down is still comfortably cool. The heat penetrates only a few centimeters during a single day. The sand cools off again during the night. Admittedly, all the sand gradually warms up during the hot season. The reverse is also observable: In a garden at midlatitudes, the freezing air temperatures during winter may last many days. Yet, the soil will freeze only down to a depth of about 40 centimeters. In general, the depth reached by a source or sink of heat is proportional to the square root of the time available for heat exchange. For instance, since freezing temperatures in midlatitudes last about a hundred times longer than a hot day, winter cold may be expected to penetrate about 10 times deeper than a day's summer heat, as observed.

The same argument applies to the solar corona. The time needed to transport heat along a loop is greater the longer the loop, proportional to the square of the loop length. Correspondingly, the density in a longer loop can be somewhat smaller so that the heat is radiated away somewhat more slowly. There is some confirmation that this relation really works for the Sun. Loops expected to be longer on the basis of their gas density and temperature are actually longer. However, the confirmation is weak because the range of observed loop lengths is limited. At the very most, they span half a solar radius. At the opposite extreme, they merge into the X-ray bright points, which do not fit this theory because they radiate their heat right where it is generated.

Density Adjustment

The skeptic will not yet be satisfied. Why should the gas density ever assume the coronal value? Why not a hundred times higher or lower? Or a thousand times? Why is it different inside and outside X-ray loops?

How is the density adjusted? Imagine a loop in which energy is supplied steadily at the top, conducted downward, and radiated away from the feet. Now imagine that the rate of energy supply at its top is suddenly doubled. The extra energy cannot be radiated away at the top. It cannot be radiated away even if it travels to the feet of the loop, because the density there is adjusted so that only the normal heat supply will be radiated away. The excess energy must travel further down, into the transition zone and perhaps into the chromosphere, where it may indeed be radiated away. However, the excess energy suddenly arriving also heats the gases there. The heating causes the gases to expand. They cannot expand downward toward the much denser photosphere. Following the path of the least resistance, the gas rises. As a result, more gas enters the corona and is heated to coronal temperatures. The density in the loop increases. This process continues until the density at the foot of the loop is sufficient for the gas to radiate away even the increased heat supplied from the top. Then the loop once again attains a new steady equilibrium. This scenario has been supported by some detailed calculations. Progress is slow because the calculations require much complicated atomic physics, even though the basic idea is quite simple.

We can therefore conclude that the density of gas in a coronal loop is determined by the energy supplied to the top of the loop. The denser, more easily observable loops are the ones with a higher rate of energy supply.

What have we learned? We have merely replaced one question (How is the density adjusted?) with another question (How is the energy supplied to the loop?). One way the corona may be heated is by waves traveling upward from the photosphere, but these would not be ordinary sound waves.

Coronal Heating by Alfvén Waves

The magnetization of the corona allows several forms of waves to develop in addition to ordinary sound waves. The wave of special interest is called an Alfvén wave. It occurs because magnetic field lines are somewhat analogous to taut strings. Anything under tension can carry waves. A taut string, struck by a sideways motion, carries waves along it. In the case of a violin string, the waves on the string are transformed to waves in the air that we hear as music. Similarly, magnetic field lines struck by a sideways motion carry waves along the field lines (see figure 8.2).

The waves travel at what is called the Alfvén speed. It is proportional to the strength of the magnetic field and also increases as the gas density decreases. In most of the corona, the Alfvén

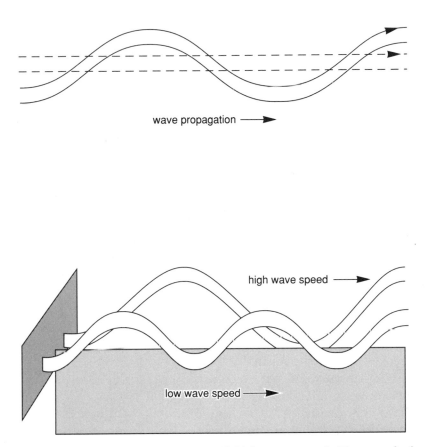

wave propagation ⟶

high wave speed ⟶

low wave speed ⟶

Figure 8.2. Alfvén waves. Magnetic field lines act much like stretched strings. When sideways motion distorts them into a wave pattern, this pattern travels along the field lines with the Alfvén speed. Top: Alfvén wave in a homogeneous medium. Bottom: Alfvén waves on either side of a boundary across which the Alfvén speed changes. A typical boundary is the edge of a coronal loop, where the gas density is larger and the Alfvén speed less inside than outside. Each wave progresses at its own Alfvén speed. Where the two adjacent waves are out of phase, localized currents are created that can be turned into heat.

speed is about 1,000 kilometers per second, which is several times the coronal speed of sound.

Alfvén waves have an interesting property. They travel along magnetic field lines and follow the field lines even if these are slightly curved. Consequently, strong Alfvén waves may travel along certain coronal magnetic flux tubes whereas much weaker waves travel along neighboring flux tubes. If stronger waves release more energy and thereby bring more gas into the loop, then a magnetic flux tube with stronger Alfvén waves will become observable as a coronal loop emitting X rays while neighboring flux tubes remain unobservable. Therefore, Alfvén-wave heating naturally yields a magnetically structured corona.

Alfvén waves can travel long distances without being dissipated. At first, this made Alfvén waves appear very suitable for coronal heating because they could successfully reach the tops of coronal loops without being dissipated further down. However, the Alfvén waves tend to travel too far. They do not readily deposit their energy as heat. Heating requires a breakdown of the frozen-in approximation (see chapter 7); that is, it requires a situation in which electrical currents are very localized. Fortunately, Alfvén waves traveling along the surface of a coronal loop may fulfill this requirement. If the Alfvén speed changes across the surface of a loop, adjacent waves get out of phase, as shown in figure 8.2. Strong, localized currents are created that can indeed heat the surface of the loop.

Heating the surface of a loop raises an observational question: Why do we not see hot surfaces on loops? The answer at present is not very satisfactory: The hot surfaces are so thin and contain so little gas that they emit unobservably little radiation. The next generation of space experiments is anxiously awaited.

An alternative way in which the corona might be heated would be through a large number of "microflares," as described in chapter 9. Neither of these theories is really satisfactory. In particular, one cannot yet explain why any one magnetic flux tube is heated and filled with gas while a neighboring one is not. Consequently, the gradual development of active regions requires intense study, including continual and precise measurements, for years to come.

Leap to the Stars

Perhaps the solar corona is "predictable" simply because coronal heating has a limited range. If so, the theory is not nearly as attractive because it then depends on the specifics of coronal heating. Accordingly, other stars might have drastically different heating and X-ray emission rates.

A test based on observations of other stars is possible. When launched in 1978, the satellite named *Einstein* carried equipment to measure stellar X rays. *Einstein* data not only proved that other stars have coronae, but they provided X-ray fluxes and coronal temperatures for several hundred stars. Some of these stars are very active. Others are rather ordinary and relatively quiescent.

The question is, Are the coronas of the ordinary stars similar to the corona of the Sun? One point is encouraging. The observed coronal temperatures of the ordinary stars are in the range from 1 to 5 million degrees, which is in excellent agreement with the range observed on the Sun.

The *Einstein* data provide only the X-ray flux from an entire stellar corona. There is no direct observational information on the density or the heating rate of any individual loop. No complete test of the theory is available. But one may use the theory to derive a loop density and then to ask whether the results make sense.

One adopted "recipe" specifies that the loop length equals the height of the stellar corona at which the gas density is half the value at the coronal base. This value is about 40,000 kilometers for the Sun. It is larger for stars with a lower gravity or a hotter corona. The gravity of stars is known from other astronomical measurements. The coronal temperature can be obtained from the *Einstein* measurements. Therefore, the hypothesis states, we know the loop length on every star measured by *Einstein*. The observed coronal temperature and the deduced loop length, plus the theory of conduction-dominated loops, allow us to predict the gas density of the loops on that star and the X-ray emission from any parts of that star covered by coronal loops. Finally, comparison with the observed total X-ray flux indicates what fraction of that star is covered by coronal loops.

If many ordinary stars appear to have a reasonable fraction of their surfaces covered by coronal loops, then we have some support for both the hypothesis concerning the loop length and the theory of conduction-dominated loops. However, if the stars appear to be more than totally covered by loops, then either the hypothesis or the theory, or both, are invalid.

When this prescription is applied to the Sun, as if observed from afar, roughly 10 percent of the Sun is predicted to be covered by coronal loops. The value is reasonable and allows some confidence in using the prescription for other stars. But some caution is in order. The coverage of 10 percent is predicted for both sunspot minimum and maximum, contrary to observation. No change in coverage is recognized because the change in X-ray emission between minimum and maximum can be explained totally by a change in coronal temperature. This result provides an indication of the likely errors in this procedure.

Einstein Results

The results for other stars fall into two classes. Stars with a photospheric temperature hotter than 10,000 degrees either have no corona or they have a much more intensely emitting corona. The theory based on the solar corona does not seem to apply there.

Among the cooler stars, the predicted coverage is reasonable, generally roughly 10 percent, and never more than 100 percent. Therefore, the *Einstein* observations provide some support for the solar theory of conduction-dominated loops. These observations also provide one final argument against coronal heating by shocks: The theory for shock heating predicted stellar coronae only for a narrow range in stellar surface temperatures, contrary to observation.

Theorists find conduction-dominated magnetic loops attractive because the theory involves mostly basic physics and remarkably little astronomical detail. We should remember, however, that the solar corona does not consist merely of loops each 40,000 kilometers long. It also has X-ray bright points, which do not fit the theory. Probably X-ray bright points exist on other stars also, but they cannot yet be observed.

Chapter 9

The Combed Corona

Space Advantages

The brightest, innermost parts of the corona are photographed routinely and easily from ground-based coronagraphs. Regions somewhat further out can be photographed during an eclipse, but the extra distance is quite limited. Our atmosphere retains some residual sky brightness even during eclipse, because the atmospherc adjacent to the path of the eclipse remains sunlit and scatters sunlight into the eclipse zone. Consequently, the faintest parts of the corona are still obscured. The longer the eclipse, the wider the eclipse path, the darker the sky, and the better one can observe faint phenomena. Eclipse expeditions seek out the eclipses with the longest duration, not only because there is more time to observe, but also because the sky is darkest during these eclipses. Nevertheless, space-based cameras can photograph coronal features with extremely low brightness much further out from the Sun than is possible from the ground.

Coronagraph

Skylab carried a camera called the white-light coronagraph. Figure 9.1 shows one photograph taken with this instrument. The optics within the camera are arranged so that the outermost, faintest sections are enhanced in comparison with the inner ones.

The outer corona in figure 9.1 consists of several streamers. They extend out to some seven solar radii from the center. Typi-

Figure 9.1. Coronal streamers, observed with the Skylab white-light coronagraph, reach at least six solar radii above the solar surface. An occulting disk blocks out both the solar disk and the innermost, brightest part of the corona, and it must be observed by other instruments. A part of the corona is obscured by the leg that supports the occulting disk. (High Altitude Observatory and National Aeronautics and Space Administration)

cal densities there are roughly 1 percent of the densities at the base of the streamers. Presumably, the streamers continue still further into space. Indirect measurements, based on solar electrons reaching the Earth, suggest that the streamers reach at least halfway to the Earth, roughly 100 solar radii from the Sun.

Long-lived Streamers

Most of the streamers in figure 9.1 hover over large active regions. Since large active regions last for many weeks, one could expect similar lifetimes for the streamers. Unfortunately, eclipse observations are separated by so many months that no streamer can be expected to be recognizable in consecutive eclipse photographs. Eclipses cannot be used to estimate the lifetime of streamers.

Skylab provided such observations. Structures such as those

in figure 9.1 come into view once every two weeks during the Sun's rotation, appearing alternately to the East and to the West of the Sun. The main structures are usually recognizable when they reappear, but details change noticeably. Bifurcations and substreamers appear or disappear. Gradually, over an interval of one or two rotations, the small changes add up so that even the main structure has altered appreciably.

A lifetime of many weeks appears to be quite normal. Of course, it has been known for some time that solar flares occasionally blast some coronal material into space and necessarily leave behind a suddenly rearranged set of streamers. But such a rearrangement was thought relatively rare.

Transients

Skylab experiments indicated that the corona changed with time not only in response to flares. Plate 5 and figure 9.2 show a phenomenon that was totally unanticipated at the time. They show a portion of the corona lifting off and rising into space for no obvious reason. No flare triggered these events. For lack of an obvious cause, these phenomena are called transient events, or simply transients.

On August 21, 1973, the erupting prominence shown in plate 5 apparently formed the lower border of the departing bubble and was carried along as the bubble rose. Perhaps the release of the prominence allowed the entire bubble to rise, but there is no obvious cause for that release either.

Since prominences and transients are observed with different equipment, to observe both, as in plate 5, is relatively rare. But the association of prominence and transient may actually be fairly common. One common feature of prominences and transients is their sudden ascent without any apparent cause. Prominences are usually quiescent while they are observable at the solar limb, yet they may suddenly take off, as in plate 5 and figure 4.5, sometimes because of a flare but often for no apparent reason at all. Prominences appearing as filaments on the disk, as in figure 7.5, may last for days and even weeks. Yet, occasionally, a dark filament on one spectroheliogram cannot be seen in

Figure 9.2. Transient observed aboard Skylab on June 10, 1973. The white-light image gives the strong impression of a gaseous bubble expanding into space. (High Altitude Observatory and National Aeronautics and Space Administration)

a spectroheliogram taken merely a few minutes later. The phenomenon has been given the French name "disparation brusque." We now know that the filament "disappears" because of its sudden and rapid ascent.

Speed of Ascent

A second similarity between prominences and transients is that they ascend at about the same rate. The transients often look like a gaseous bubble with a clearly delineated leading edge. Data from the Solar Maximum Mission show the leading edges of many transients accelerating up to a final, steady speed in the range of 200 to 1,000 kilometers per second. Roughly an hour is needed for the gas to accelerate to this speed. Similar values apply to erupting prominences. All the observations are quite consistent with the idea that magnetic forces push the bubbles and prominences outward.

Loss of Corona

How much gas is removed from the corona by one transient? In terrestrial terms, the gas weighs a billion tons, which is only the weight of a small mountain. In terms of the Sun as whole, it is tiny because the solar wind carries off that much gas every 10 minutes, and in a much less dramatic and more continuous manner. In terms of the solar corona, it is the gas normally residing in the lowest 20,000 kilometers of a coronal streamer covering an active region of about 100,000 kilometers in diameter. The transient removes a substantial portion of the corona. Some of this gas may have been part of a prominence, but most of it was originally coronal gas.

Can transients remove all of the corona? One transient removes the contents of the corona over a solar active region covering roughly a thousandth of the solar surface. When a tenth of the solar surface is covered by active regions, near solar maximum, about a hundred transients suffice to remove all the gas present in the corona at one time.

What time interval is required for a hundred transients? Skylab data probably missed many transients, because only those near the limb could be observed and because the 90-minute interval between most observations was too long compared to the rapid changes in most transients. The best possible estimate suggested one transient per day near sunspot maximum. At this rate, a hundred transients occur in about three months. Therefore, it takes merely three months to remove all of the corona into space.

A loss time of three months for the entire corona seemed like a surprisingly short time, given the apparent longevity of individual large coronal streamers. Many experts were skeptical of the various estimates that led to the time scale of three months. Accordingly, observing plans for the Solar Maximum Mission made transients an important component. One transient observed by the Solar Maximum Mission appears in figure 9.3.

The Solar Maximum Mission observed transients in much greater detail and so frequently that they could be captured in a movie. That movie reminds me of scenes at a beachfront during a hurricane when all the trees are bent by the winds, and when

Figure 9.3. Transient observed by the Solar Maximum Mission on April 14, 1980. (High Altitude Observatory and National Aeronautics and Space Administration)

the bending changes vigorously as the winds change. On the Sun, the winds are replaced by the transients. The bending trees are replaced by the radial structures in the corona. The dynamic impression given by the movie is qualitatively different from the static coronal snapshots taken by Skylab. More detailed analysis of these coronal motions leads one to estimate that the coronal gas is replaced in merely about three weeks!

Apparently the long-lived streamers have a "memory" and reconstitute lost parts. A memory is also indicated when some disappearing filaments are reconstituted after some days. The memory is presumably magnetic and resides in the underlying active region.

During 1980, the Solar Maximum Mission observed almost

one transient per day, compared to one transient per two days observed on average by Skylab. The increase at solar maximum seemed to make sense. But perhaps the high rate of transients was anomalous. In 1985, when observations by the Solar Maximum Mission resumed, the rate of transients was closer to one a week. This rate is some three times less than the Skylab observations, though both were made at the same near-maximum phase of the solar cycle. Moreover, the transients of 1985 were only about half as fast as those of 1974. Perhaps transients of 1996 will tell us what is "normal."

Combing the Corona

The lateral boundaries of some transients are observed to straighten, to become more nearly radial as the transient rises. The boundaries are thought to follow magnetic field lines. Apparently, the bounding magnetic field lines become more nearly radial. In general, all the coronal magnetic field lines experiencing transients are expected to become more nearly radial. If a hairy-ball model of the coronal magnetic field (figure 5.10) were constructed, it would appear more "combed out" after the transient than before.

The magnetic fields becoming more nearly radial due to transients constitute a magnetic simplification. New complexity is introduced by the magnetic flux emerging from the photosphere. A transient may well make room for this emerging flux. It may also alter the heating of older coronal loops and thereby change their visibility. In this sense, a transient has a very pervasive influence. Figure 9.4 shows one possible magnetic pattern before and after a transient.

The prominences are manifestations of rather high magnetic complexity. They must be supported against gravity by some form of magnetic basket. The prominence gases, being as cool as sunspots, must also be magnetically enveloped to prevent heating by conduction from the much hotter corona. The partial removal of a prominence with a transient may constitute significant magnetic simplification.

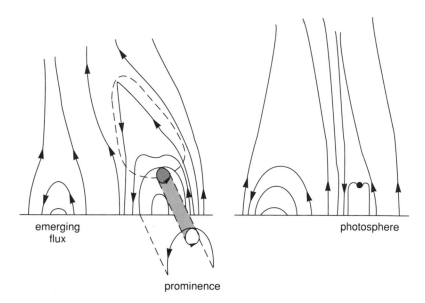

emerging
flux

photosphere

prominence

Figure 9.4. Magnetic effect of a transient. The left sketch shows a prominence embedded in a magnetic "basket" residing within an old, large magnetic pattern resembling a streamer. Newly emerging magnetic flux appears on its left. The dashed curve outlines the material that may rise as a transient. The right sketch shows the field after the transient is gone and the field lines connecting to it have been stretched out into space. In effect, the magnetic field has been straightened, "combed out." A prominence may form again in the old magnetic field. The newly emerging flux has the opportunity to occupy more space. Perhaps it adds a new prong to the old streamer.

Magnetic Disequilibrium

What causes the transients? The best explanation may be given in terms of figure 9.4. When magnetic flux first emerges, its field lines reside rather low in the corona. They respond readily to any changes in the photosphere. They are held down by the internal magnetic anchor. Days or weeks later, as the field lines arch higher into the corona and as their feet become more widely separated, they respond less readily to control from the photosphere. At some stage, control disappears. Field lines that are no longer anchored begin to rise. The energy inherent in the local electrical currents is used to push the gases upward. A transient is observed.

B. C. Low, at the High Altitude Observatory of the University of Colorado, and other theoreticians have developed mathematical solutions corresponding to this scenario. They have found a few simple but plausible solutions for slowly growing coronal magnetic loops and have shown that these solutions lose their validity when the loops become big enough. Further growth must cause disruptive motions. The transition from equilibrium to disruptive motions is quite abrupt. A transient may occur when the magnetic loops reach this size.

This attractive theory for transients involves a basic physical difficulty. The transient bubble must be disconnected from the photospheric magnetic fields. Disconnection of field lines implies a breakdown of the frozen-in approximation (see figure 7.3) and dissipation of electrical currents. Dissipation of coronal currents in a fraction of an hour requires the currents to be channeled into a cross section no wider than about 1 kilometer. One may speculate that such channeling is a natural consequence when part of a prominence starts to rise with a transient. But it has not been demonstrated to occur.

Coronal Heating by Microflares

Fortunately, the dissipation of coronal electrical currents is known to occur for another reason: Electrical currents heat the corona. There is no other plausible source of heating. Moreover, the observed magnetic control surely involves electrical currents. Therefore, the observed heating of the corona informs us that the frozen-in approximation breaks down somewhere and in some fashion. One possibility involves Alfvén waves at the surface of coronal loops (see figure 8.2). This is a very quiescent process. Theoreticians like it because it is mathematically tractable. But the observations of flares and of the sudden releases of prominences suggest a more impulsive alternative: Numerous explosive releases of heat from temporarily very localized electrical currents, essentially microflares. Observers tend to prefer this explanation for coronal heating, and it accounts more directly for the sudden release and magnetic disconnection of prominences and transients.

Guaranteed Magnetic Turmoil

How does one visualize the formation of localized electrical currents? Figure 9.4 shows several places where oppositely directed magnetic fields approach each other closely. The change in magnetic direction implies electrical currents. The more closely the opposing magnetic fields approach each other, the stronger the currents. If the fields approach to within 1 kilometer, the currents are dissipated into heat. More fields can then approach each other closely, and so on.

Will such current dissipation actually occur? The optimists argue that such situations are inevitable.

The motions at the photosphere are extremely complex. Granulation moves the field lines about. So does the supergranulation, on a larger scale (figure 7.1). All these motions are essentially random and unpredictable. All move the feet of coronal field lines, and field lines higher up must somehow adjust accordingly. Many of the motions intertwine the field lines. Many other motions twist the field lines. The resulting knots and twists are expected to run upwards and to accumulate somewhere near the top of a loop. So do the associated electrical currents. How long can this go on? Probably, the currents reach a threshold value, after which time the knots and twists interact convulsively and are converted to heat. The knots and twists disappear, the heat is conducted away, and the remaining, simplified magnetic fields are ready for another period of quiescence.

If all the electrical currents created by this turmoil are dissipated, the resulting heat is just about sufficient to heat the corona. Moreover, the upward migration of magnetic twists implies preferential heating near the tops of the loops. The theory is attractive, argue the optimists. Is there any observational support for it?

Microflares

The dissipation of localized electrical currents is a sudden event. Such seemingly impulsive releases of heat are indeed observed. They are generally called microflares. Perhaps they are merely the much smaller and much more frequent relatives of ordinary solar flares. Microflares have been recognized only in the past few years because they can be detected only by sensitive equipment.

Microflares were first detected in X-ray data as extremely tiny increases in the X-ray flux observed from a balloon flight on June 27, 1980. Whereas large X-ray flares are very rare and moderate ones may be observed daily, the X-ray microbursts may occur every few minutes. "May occur" means they occurred every few minutes during the two hours of X-ray data that have been suitably analyzed. They might not occur at different hours, days, or weeks. Are they actually small flares? The observing instrument cannot tell where in the corona the microbursts originate. However, sudden changes have been detected in tiny regions of spectroheliograms coinciding in time with the X-ray microbursts. Therefore, the X-ray microbursts are a possible indication of coronal heating by microflares.

Microbursts have also been detected at certain radio frequencies. Careful analysis indicates that the radio microbursts have properties similar to those of stronger radio bursts observed with flares. This is a second, somewhat indirect indication of impulsive phenomena in the corona.

In addition to these microbursts, there is a kind of radio burst that is not associated with flares and has been observed for years. A typical burst lasts for less than a second. A dozen bursts per minute is not unusual. The bursts are emitted from anywhere in the corona over a developing active region. I interpret this type of radio burst as additional evidence for impulsive heating and microflares in the corona. However, other interpretations also exist.

Coronal Heating and Transients

Is coronal heating actually due to microflares? What about the alternative involving Alfvén waves? Both theories have some points in favor, some against. Both make some predictions that need to be checked by the next generation of space experiments. The ultimate explanation may well incorporate aspects of both theories.

If coronal heating is due to microflares, magnetic simplification and current dissipation are frequent and closely related events. The sudden release of a prominence or of a transient then merely represents one of the largest and relatively rare magnetic-simplification events, just as a major flare represents one of the largest and relatively rare current-dissipation events.

Chapter 10

Coronal Holes and the Solar Wind

Coronal Hole Number One

Images of the Sun in soft X rays sometimes show distinct regions of very low emission, dubbed coronal holes. Plate 12 shows the most prominent of about half a dozen coronal holes recorded by Skylab. It is variously called "coronal hole number one" or the "boot of Italy" hole, since its outline for some time looked like a map of Italy.

Coronal hole number one was not only one of the largest coronal holes but the one that extended farthest in the north–south direction. After it was first observed, solar astronomers expected that it would soon be highly distorted in shape because the equatorial parts would rotate more rapidly than the more nearly polar parts (figure 7.4). The distortion should have been sufficient to make the hole unrecognizable after only four rotations. Nothing like that happened. This coronal hole was observed for at least eight rotations. The hole suffered merely some minor distortion.

Magnetic Neglect

With time, the hole was gradually filled in from the edges. The neighboring coronal structure "encroached like crabgrass into the hole," as visualized by J. Eddy.

Apparently, coronal holes occur where there are no nearby active regions. Skylab observations took place when active regions were of rather modest extent. Skylab recorded several coronal holes, including the large coronal hole number one. The Solar Maximum Mission made its observations in 1980 when there were many active regions, and recorded only a few small holes. It is useful to think of a coronal hole as a region of benign magnetic neglect.

Why Dark?

The coronal hole is observed as a lack of radiation because the gas density there is very low. Why is the gas density different there? The photosphere under a coronal hole looks normal. The spicules look normal. Apparently, the energy available to heat the corona is normal. Something important must be different at coronal heights. Somehow the gas in a coronal hole must respond differently to the heat supplied from below than does coronal gas over an active region.

The difference resides in the shape of the magnetic field lines. They are "closed" over active regions: We speak of coronal loops and imagine field lines emerging from the photosphere at one foot of a loop and reentering at the other (figure 8.1). But the field lines are "open" over coronal holes: They emerge from the photosphere and continue outward. They must turn around somewhere, but beyond the region of observation.

The open nature of the magnetic field lines can be demonstrated by a "hairy-ball" computer model, as in figure 10.1. The computer has eliminated the ordinary coronal loops that normally dominate models such as figure 5.10 and has included only field lines that reach at least 0.6 solar radii above the surface. The coronal hole appears like a "part in the hair." Most of the field lines reaching interplanetary space come from the coronal hole.

Open magnetic fields allow gas to flow from the Sun toward interplanetary space. The coronal hole is empty simply because the gas has been driven outward. The resulting solar wind encompasses the Earth and is reponsible for important solar-terrestrial interactions. It reaches beyond the planets, merging with the interstellar gases at a distance estimated to be several times the distance to Neptune.

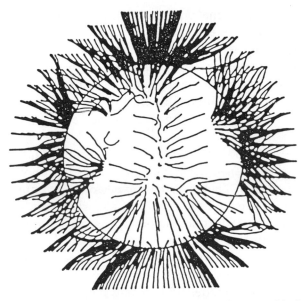

Figure 10.1. Open magnetic fields associated with coronal hole number one on August 21, 1973. For this "hairy-ball" model of solar magnetic field lines, only field lines reaching at least 0.6 solar radii above the surface have been retained. The "part in the hair" represents field lines reaching interplanetary space. They are the distinguishing features of the coronal hole. Field lines near the limb belong to high coronal streamers. (Courtesy Randolph H. Levine and *Astrophysical Journal*)

The coronal holes may be dark, but, as the source of the solar wind, they are far from uninteresting. This role was not appreciated until the Skylab observations were analyzed.

Solar Wind Prediction

Eugene Parker, of the University of Chicago, theoretically predicted a steady solar wind as early as 1956, years before any space probe could check the prediction. Parker argued that the gas pressure in the solar corona is great enough to drive coronal gas away from the Sun, despite the solar gravity, and that the resulting steady solar wind would reach supersonic speeds. His argument was prophetic on two counts. First, the corona is dynamic. No static corona is possible, despite appearances during eclipse. Motions are unavoidable. Second, the wind is supersonic. Parker argued that only a supersonic wind could carry enough pressure to merge reasonably with the distant interstellar gases. Any model involving merely a slow coronal evaporation would yield an implausibly limited corona. The interstellar gases would encroach upon the planets, which is an implausible situation.

Parker estimated that the wind passes the Earth at a few times the coronal speed of sound, specifically, at about 400 kilometers per second. The first space experiment to enter the solar wind not only detected the wind but also showed a wind speed close to the predicted value. Theory appeared triumphant.

The first wind observation was somewhat fortuitous. Later observations frequently showed wind speeds up to about 700 rather than 400 kilometers per second. Parker's theory could not accommodate such values. Parker had made an assumption essential for an inital mathematical solution, namely, that the wind and its source at the Sun are unmagnetized and spherically symmetric. Was that assumption wrong?

Source of Wind: Coronal Holes

Skylab demonstrated the importance of magnetization. Contrary to expectation, the solar wind does not come from the hottest and most active parts of the corona, because these are magnetically confined. The wind must arise in the magnetically open portions of the corona, namely, the coronal holes.

Skylab observations clearly associated coronal hole number one with high wind speeds. The hole, rotating with the Sun, reappeared about once a month. Fast winds typically arrived at Earth three or four days after the coronal hole faced the Earth. Since three or four days is a typical transit time for a wind traveling radially from the Sun to the Earth, these observations are consistent with the fast wind emerging from the coronal hole. The hypothesis could be confirmed in detail. Given the observed wind speed at any time and the radial direction of the wind, one could compute just where on the Sun that portion of

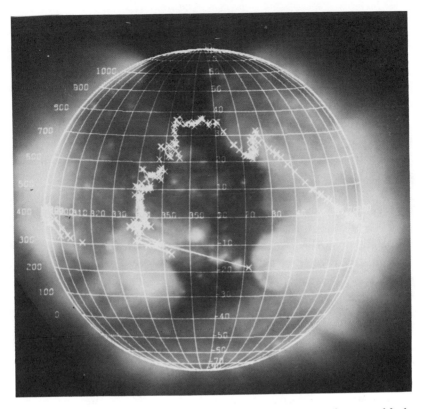

Figure 10.2. Source of solar wind. The Sun is shown with coronal hole number one, superposed by a plot of the speed of the solar wind against source longitude on the Sun. The source is plotted at that longitude that must have faced the Earth when the wind left the Sun, assuming that the wind travels radially and steadily at the speed observed at Earth. The plot demonstrates that the fast wind originates in the coronal hole. (Courtesy L. Golub and National Aeronautics and Space Administration)

the wind originated. Figure 10.2 shows the detailed results. Clearly, the fastest wind originates in the coronal hole.

Fast Winds: Magnetic Control

A coronal hole seen at eclipse, as in figure 4.2, looks like a megaphone. It widens substantially with height. Its borders are thought to act like a magnetic channel. Gas flow through this rapidly diverging channel is quite different from the radial flow Parker had assumed. Indeed, the divergent channel causes very efficient wind acceleration. As the gases rise, they must thin out by sideways expansion to fill the available funnel. With this sideways expansion, the gas pressure decreases. The gas pressure decreases outward much more rapidly than in Parker's theory. This more sudden drop in pressure, in turn, creates a faster wind. The channel widens enough to explain winds up to about 700 kilometers per second.

The megaphone-shape of coronal holes has a second consequence, which may be even more significant. It permits many possible mathematical solutions for the steady wind. These involve shocks within the funnel where the wind is accelerating. Within any given period of a few hours, the wind presumably obeys one of these solutions. Which one? The correct solution depends on the previous gradual changes in the wind. It may depend on the way the wind originated many days earlier when the coronal hole formed in response to the retreat of some aging active regions. This slow formation cannot yet be treated theoretically. Consequently, one also cannot evaluate just how the wind is accelerated within a coronal hole. Such ignorance leaves the theory for the solar wind incomplete. More clues must be sought in the observations.

Slow Winds

If high-speed winds reach us from coronal holes, what is the source of the winds with the lower speeds, the speeds that were easily explained by Parker's theory? Figure 10.2 suggests that this wind arises from the corona over active regions. How can wind emerge from magnetically closed regions? Perhaps they are not really quite closed. Perhaps individual magnetic flux tubes tunnel through the coronal streamers. Such flux tubes would be quite unobservable. The assumptions made to construct a "hairy-ball" model of the coronal magnetic fields preclude the appearance of such flux tubes in the model.

One might expect such flux tubes on geometrical grounds:

There are "parts" in every hairy-ball model, such as that of figure 5.10. The field lines emerging near the "parts" have the opportunity to reach far above the surface. How far cannot be determined by the model.

One might also expect such flux tubes to exist on physical grounds: Holes may occasionally be punched into the closed magnet fields. For instance, G. Brueckner of the Naval Research Laboratory and his colleagues photographed the Sun from rockets and afterward recovered the payload and the exposed film. They obtained an unusually high spatial resolution of about 500 kilometers on the Sun, which is far superior to that from satellite data. The images show places on the Sun where gases rise at high speed. The fastest of these "jets" are no wider than about 2,000 kilometers and rise at speeds up to 400 kilometers per second. Brueckner interprets the jets as exploding tiny magnetic loops. They may travel upward somewhat like bullets, punching through the coronal magnetic fields. The solar community is somewhat skeptical of such suggestions and counters that the jets may just be relatively rare, very energetic versions of the common spicules. The latter are widely thought to cause nothing more than shocks that heat the lowest part of the corona. Nevertheless, the transition region is in almost continual rapid motion. No one can really say what happens to these motions, to the spicules, and to Brueckner's jets. Once again, observations are needed from the next generation of space experiments.

Ultimately, all of the solar wind is determined by the magnetic geometry at the solar surface. A substantially better measurement of the magnetic fields and currents at the surface would help us to understand the source and the observed variations in the solar wind. Much of the effort that is currently directed at improving ground-based equipment includes such measurements.

Wind Speed Data

It took five years after Sputnik's launching in 1957 to confirm the solar wind observationally. Since then, most of the data about the solar wind near the Earth were acquired by a series of probes called Interplanetary Monitoring Platforms, and later by three craft called International Sun–Earth Explorers. Two European spacecraft, *Helios I* and *II*, explored the wind about halfway to the Sun. The Pioneer and Voyager spacecraft, having visited Jupiter and further planets, are now exploring the solar wind in the outer solar system. Much interesting detail has been learned about the solar wind. However, it has been difficult to explain the observed wind speeds and densities at the same time. The discrepancies have spurred many refinements in the theory. Agreement between theory and observations has improved, but has not yet reached a satisfactory level. One problem is that the wind near Earth originates at fairly high solar latitudes. We know very little about the three-dimensional structure of the solar wind. Much new information is to be obtained by the space probe *Ulysses*, originally named International Solar Polar Mission, when is passes over the poles of the Sun sometime in the 1990s.

Fortunately, there are two other ways to observe the wind near the Sun. They are less direct than data from *Ulysses*, but they may be available much earlier and over a longer period of years. Both methods show initial promise.

One measure of the solar wind requires the electronics technology developed for radar. Figure 10.3 shows a radio antenna at Earth receiving radio signals from a Voyager space probe at a time when the line from Voyager to Earth passes near the Sun. At that time, the radio waves must pass through the solar wind close to the Sun. The wind carries with it small density fluctuations that alter the received signals in minute but measurable ways. The measurements inform us about several properties of the wind, including its speed. So far, the closest distance to the Sun measured in this fashion has been at a heliocentric distance of 1.7 solar radii, by the space probe *Helios I*. The derived wind speeds are reasonable, but they cannot yet distinguish among the various models for the wind. One must await a radio source with a line of sight that comes close to the Sun frequently or for a more extended period of time. Some cosmic radio sources do that, but their measurement involves additional difficulties that have not yet been addressed.

A second measure of the wind makes use of the very tiny fraction of neutral hydrogen atoms in the corona. A few such

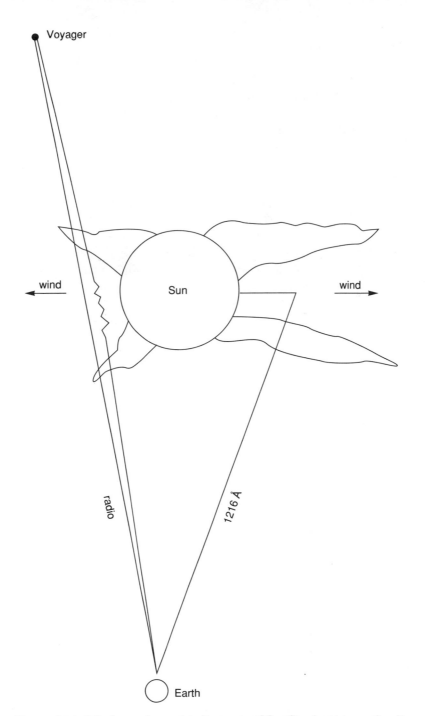

Figure 10.3. Wind speed may be determined by the alterations of radio signals passing through the corona, for instance, from the Voyager spacecraft, or by the intensity of scattered atomic ultraviolet radiation such as HI 1216 Å, which weakens for a faster wind.

neutral atoms persist in the corona despite its high temperature. Just as the coronal electrons scatter white sunlight toward us, the neutral atoms scatter radiation at specific atomic wavelengths. The geometry is shown in figure 10.3. A typical atomic wavelength is that of neutral hydrogen at 1216 Å (see figure 4.6). The amount of light scattered toward us depends on the motions of the atoms that scatter the light. If they are nearly at rest relative to the Sun, they scatter fairly efficiently. If they move rapidly toward or away from the Sun, they do not scatter at all. Consequently, measurements of the intensity of the scattered radiation tell us of the velocities of the atoms toward or away from the Sun. The wavelengths used must be in the ultraviolet, to avoid the bright white light from the disk, so that the measurements must be made from space. Existing measurements are not yet sufficient to resolve the outstanding problems concerning the solar wind.

Comet Tails

The solar wind was first anticipated because of its effect on comet tails. Comet tails can be classified into two types, as vividly demonstrated in figure 10.4. One type consists of dust. The dust absorbs solar radiation and is blown away from the Sun by the pressure of the radiation. Dust tails tend to be somewhat curved. The other type of tail consists of various ionized molecules. These tails are more nearly straight. They are blown away from the Sun more efficiently than the dust. However, the radiation pressure is much less efficient. What accelerates the molecules? L. Biermann, who had earlier formulated the theory for coronal heating by shocks, suggested in 1950 that the cometary molecules are picked up by friction with the gas of a solar wind, so that the molecules are quickly accelerated to the speed of the wind. The suggestion led Parker to develop his theory for the solar wind.

Just how quickly can the solar wind pick up the cometary molecules? With time, we have learned that it does so very efficiently. In fact, the acceleration occurs so close to the comet that details cannot be resolved from Earth.

Figure 10.4. Comet Arend-Roland, May 5, 1957. The straight tail consists of molecular ions that are picked up efficiently by the solar wind. The curved tail consists of dust grains accelerated more leisurely by solar radiation pressure. The arrow marks the moving remains of an explosion at the nucleus of the comet, which is overexposed and marked by a small dot. Rings around bright stars are camera effects. (Courtesy Freeman Miller, University of Michigan)

The first space measurement of this process occurred when *International Cometary Explorer (ICE)* passed comet Giacobini-Zinner on September 11, 1985. The spacecraft had been in space near Earth for several years, as the third International Sun-Earth Explorer, and had been designed with rather different experiments in mind. Its capabilities for taking measurements near the comet were limited. Nevertheless, *ICE* transmitted data about a region with strong electrical fluctuations, located only a few thousand kilometers from the comet. The fluctuations confirmed the very local merging of solar wind and cometary gases.

Similar phenomena were observed in much greater detail when the European spacecraft *Giotto* and the two Russian probes *Vega 1* and *Vega 2* passed near comet Halley in March 1986. The observations were more difficult to interpret than those of *ICE* because gas and dust are ejected much more explosively by comet Halley. Nevertheless, the plasma observations can all be interpreted to support Biermann's original suggestion. The cometary gases are picked up by the solar wind, but very much more efficiently than Biermann could have expected.

The Earth's Magnetosphere

Comet Halley passed very near the Earth in 1910. Many people feared its "poisonous gases," cyanogen (CN). Would these gases reach Earth and enter our atmophere? Now we know they cannot. Not even the solar wind can reach the atmosphere. We are protected by a cavity controlled by the Earth's magnetic field, appropriately called the magnetosphere (figure 10.5). The solar wind arrives from the left with much momentum. It is stopped at a bow shock by the pressure of the Earth's magnetic field. The bow shock occurs at a distance of about 10 Earth radii, adjusting in accordance with the changing speed and density of the wind. Some of the consequences of the impact of the solar wind on the magnetosphere are the Van Allen belts and disturbances in electric transmission lines (more in chapter 14). The geomagnetic cavity extends far to the right of figure 10.5 in what is often called a teardrop-shaped tail. As wind speed and pressure change, so does the shape of the geomagnetic cavity. The geomagnetic tail may flutter in the wind much as a flag does.

Terrestrial Effects of the Solar Dynamo

Most magnetic field lines embedded in the solar wind simply sweep around and past the Earth, just as the gas does. But some arriving interplanetary field lines are nearly antiparallel to the geomagnetic field lines at the bow shock. When they meet, strong local currents are created. The currents travel toward the Earth along the geomagnetic field lines, especially near the magnetic poles, altering the normally occurring flows in the auroral zones. These geomagnetic effects, determined by the direction of the interplanetary magnetic field, ultimately

Figure 10.5. Geomagnetic cavity protects us from the solar wind. The wind is stopped at the bow shock and sweeps around the cavity. Aurorae (northern or southern lights) occur where the magnetic funnels reach the atmosphere near the Earth's magnetic poles. The Van Allen belts are doughnut-shaped regions containing very fast electrons and protons. The geomagnetic cavity extends far downstream, forming a magnetized tail that flaps in the solar wind. (National Aeronautics and Space Administration)

depend on the direction of the magnetic field at the source of the wind—the Sun.

Observationally, most of the solar detail (such as active regions) has no influence on the wind's magnetic field. Only the largest-scale solar magnetic field matters, the field we would measure if we observed the Sun from the distance of another star. This field changes only gradually, on a time scale of many months.

During 1982 the large-scale magnetic field looked somewhat like the Earth's magnetic field, as if a bar magnet were located at the solar center (see figure 5.5). However, the solar magnetic north–south axis was tilted by about 50 degrees relative to the axis of rotation. Accordingly, the magnetic equator was a great circle tilted 50 degrees relative to the rotation equator. Differential rotation would have distorted that great circle in short order. It did not. Apparently, the large-scale solar magnetic field is anchored deep in the Sun, beyond the influence of surface differential rotation. The magnetic anchor is associated with the solar dynamo. The fact that the solar wind brings to us only this large-scale, deeply anchored field is consistent with the notion that the source of the wind is the coronal holes, which also experience no differential rotation.

Seen from the distance of another star, the solar magnetic field of 1982 would have changed direction twice during each solar rotation. It would have been directed toward the observer for about two weeks, then away for about two weeks, and so on. The solar magnetic pattern of 1982 was unusually simple. In most years, the pattern is more complicated. Seen from afar, the magnetic direction would appear to reverse four times per solar rotation. Just as the resulting interplanetary magnetic direction changes two or four times per month, so do the observable geomagnetic effects.

The large-scale solar magnetic field reverses with every 11-year solar cycle. Its 22-year cycle may manifest itself in a 22-year cycle in terrestrial climate (see chapter 14).

We on Earth experience the effects of the dynamo anchored deep within the Sun, a dynamo of which we are still profoundly ignorant.

 PART III

SOLAR FLARES

Chapter 11

Solar Flares

White Light

Richard C. Carrington observed a surprising event in 1859. He was recording the positions of sunspots and suddenly observed a brightening near one of the spots. The brightening had the shape of two ribbons running across the umbra of the spot. The area brightened and faded again in the course of some 15 minutes.

Carrington had observed one of the rarest kind of flares, now called white-light flares. They are the largest of flares, occurring at most a few times in a solar cycle.

Hydrogen Light

More specialized images of solar flares became available in the 1930s. An unusually large flare of 1972 appears in figure 1 and its close-up in figure 11.1. Each figure presents a hydrogen alpha spectroheliogram, which is a view of the Sun based on the light from hydrogen in the chromosphere much like figures 4.4 and 5.11. Whereas those two pictures would have looked the same a few minutes earlier or a few hours later, the flare in figures 1 and 11.1 brightened in just a few minutes and disappeared in a few hours.

Flares are traditionally classified according to their appearance in spectroheliograms, mainly in relation to their area. The classification indicates the general importance of a flare for many other observations and for terrestrial effects.

Figure 11.1. Solar flare observed in the light of hydrogen on August 7, 1972, from Big Bear Solar Observatory, a close-up of figure 1. This flare had unusually strong terrestrial effects such as power blackouts. It produced enough radiation to kill Apollo astronauts if an Apollo craft had been in space at that time. Characteristically, it surrounds a prominent spot in the shape of a pair of ribbons. (Big Bear Solar Observatory)

Most flares are small. They cover a small fraction of an active region, comparable in size to all the other, normal structural features. They may last for a few minutes. A dozen such flares may occur in any active region in a day. If one imagines a movie made of successive spectroheliograms, the small flares constitute a rather minor flickering in the active regions. They are quite inconsequential.

A Great Flare

Great interest resides in the largest flares, such as the one in figures 1 and 11.1. Great flares may cover much of an active region and last over an hour. Often they start near or over a sunspot. Then they lengthen into "ribbons" winding along the border between regions of opposite magnetic polarity. The choice of the word "ribbons" reflects the two-dimensional view of a flare to which observers were limited before the space age. The longer ribbon in figure 11.1 reminds many viewers of a sea horse. The crispness of the picture is typical of photographs from Big Bear Solar Observatory, a site that offers a quiet atmosphere over a lake in the mountains east of Los Angeles (figure 11.2).

Frequently the ribbons light up in merely a minute or two. Classically, this interval was called the flash phase. Since the advent of space observations, it is more usually called the "impulsive" phase. It is followed by the "gradual" phase, when the flare may expand further and may cover up to 1 percent of the solar surface. Maximum area might be attained after some 10 minutes. The flare then shrinks and fades even more gradually. The last pieces may still be recognizable several hours after the impulsive phase.

Flares may change rapidly. Small flares can appear and disappear in a few minutes. Large flares can achieve maximum brightness and change their shape drastically within five minutes. Ideally, to record flares in sufficient detail one should observe the Sun continuously. A practical flare patrol consists of hydrogen alpha spectroheliograms taken automatically at fixed intervals of time, speeded up when a flare is detected. The automatic rate achieved in the 1950s, one picture every ten minutes, was still so slow that important information was missed. Now spectroheliograms can be taken every minute of the day. This makes it possible to record most of the flare development without the intervention of an observer. To analyze that many photographs would have been rather a large undertaking in the days when photographs were visually inspected on microfilm and any interesting features were sketched with pencil and paper. By 1980, when the launch of the Solar Maximum Mission

Figure 11.2. Big Bear Solar Observatory in California, operated by H. Zirin of the California Institute of Technology. The lake surrounding the observatory stabilizes the atmosphere so that the best Big Bear pictures are among the sharpest ground-based solar pictures produced without computer-based rectification. (Big Bear Solar Observatory)

prompted special efforts, some flares were photographed once every two seconds.

Routine photography of the Sun should not be interrupted by the day–night cycle. Most observatories take their best pictures in the morning before the air becomes turbulent, during an interval of only a few hours or less. The need for continuous, quality observation has led to close collaboration among observatories in all parts of the world. Fortunately, only modest equipment is needed and can be accommodated even by institutions in small and/or developing countries. Indeed, the developing countries have the advantage of lower labor costs.

In the United States, the Solar Optical Observatory Network is run by the Air Force. The Air Force is greatly interested in flares since they affect methods of radio communications and alter the surroundings, on Earth and in space, where the Air Force operates. Not only do we need to study the flares themselves but we must also be able to predict when significant flares are likely to occur (more in chapter 14).

Explosion

What are flares? Intuition suggests some sort of explosion. Figure 11.3 shows a spectroheliogram providing a strong three-dimensional sensation of an explosive event. The impression is quite correct. When ejecta speeds can be measured, they are clocked at several hundred kilometers per second, much like the speeds of rising prominences and transients (plate 5 and figures 4.5, 9.2, and 9.3). Such speeds should cause drastic heating and other violent disturbances. Is there more to be observed? Hydrogen alpha observations of flares can be compared to the experience of the blind man feeling the trunk of an elephant. The persistent blind man may seek further and find more than the trunk.

Flare Energy

A flare must indeed be more than merely an explosion in the chromosphere. The gradual phase of a flare lasts too long to be merely the radiative loss of energy released at the initial explosion. Moreover, the energy in the chromospheric explosion appears far too small to cause the known terrestrial effects. Carrington noted a geomagnetic disturbance two days after he observed the white-light flare. Further observation of geomagnetic effects made the solar connection quite evident. In the 1960s, the solar wind became an established concept and so did flare-generated

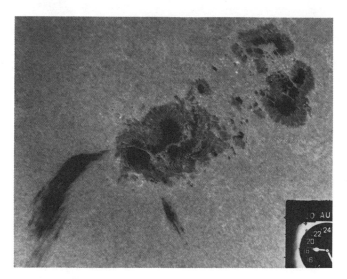

Figure 11.3. Solar flare of August 20, 1971, observed from Big Bear Solar Observatory. The dark streak yields the strong and correct impression of ejection of gases from an explosion. (Big Bear Solar Observatory)

shock waves traveling through the wind from flares to the Earth. These shocks also appeared to carry surprisingly large energies, much larger than the energies in the chromospheric explosions.

Experiments during the 1950s and 1960s led gradually to the realization that "the flare" occurs in the corona and that the chromospheric flare is merely a sideshow.

Radio Observations

A new era in flare research dawned in the 1950s after it was found that flares emit detectable radio waves. One of the most effective instruments for measuring the radio emission of the Sun was the radio heliograph built in Culgoora, Australia, in 1968. (It was decommissioned in 1985.) Figure 11.4 shows this arrangement of 96 separate radio antennas, arranged in a large circle. Together, the antennas acted like one very large radio telescope, able to pinpoint solar radio sources accurate to roughly a tenth of the solar radius. All the receivers would be tuned to the same radio frequency, a frequency of radio waves

Figure 11.4. Radio heliograph in Culgoora, Australia. The 96 radio antennas arranged in a circle of 3-kilometer diameter could map the solar radio emission, at selected radio frequencies, as often as once every few seconds. (Copyright of CSIRO Division of Radiophysics, Australia)

arising in the corona. A map of all the sources observed at this frequency at one time amounts to a radio spectroheliogram, mapping whatever coronal gases radiate at that frequency. The Culgoora radio heliograph could take a radio snapshot of the Sun every few seconds.

The radio heliograph recorded radio emission from many

flares. Typically, the emission came from coronal regions that covered the active region containing the flare. Sometimes this radiating volume appeared to move outward from the active region into space. Figure 11.5 shows "Westward Ho!," an event observed soon after the radio heliograph began operations.

The Culgoora radio heliograph was the most dramatic of the new radio installations because it could create pictures of radio sources. But it could do so for three radio frequencies at most—typically, the frequencies of 327, 180, or 80 megahertz (1 megahertz = 1 million cycles per second; for comparison, the FM radio band in the United States spans frequencies from 88 to 110 megahertz). Clark Lake Radio Observatory (plate 13) was built to map the Sun at any frequency between 125 megahertz and the lowest frequency normally observable from the ground, 15 megahertz. Clark Lake data have provided insight into phenomena in the outer corona that are difficult to observe by any other means. Such phenomena include the radio microbursts that may be indicative of frequent localized coronal heating (see chapter 9). (Clark Lake has also been decommissioned.)

However, solar radio astronomy made many advances well before the Culgoora and Clark Lake radio observatories became active, thanks to smaller radio telescopes, carefully instrumented to measure radio emissions quickly at many radio frequencies. These radio telescopes could not pinpoint the source of the radiation on the Sun, but this position was simply assumed to match the flare observed in hydrogen alpha.

Figure 11.6 shows a dramatic example of Australian radio observations in 1963. The horizontal axis measures time and spans about 25 minutes. White areas indicate the radio frequencies at which emission is intense at any one time. Low radio frequencies are at the top, high ones at the bottom. Any one receiver, measuring emission in a small range of radio frequencies, detects a series of bursts. Such a "dynamic spectrum" is a convenient way to display the data from receivers at many frequencies.

Why are the lower frequencies displayed near the top of figure 11.6? The frequencies provide a way to map height in the corona: High frequencies arise low in the corona, and low frequencies arise high in the corona. Consequently, figure 11.6 presents

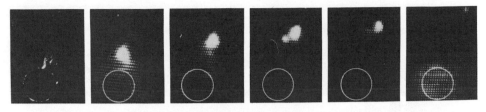

Figure 11.5. "Westward Ho!" An ejection on March 2, 1969, first observed in hydrogen alpha, was traced far out into space by its radio emission. Displacement by three solar radii in a few hours implies a speed of about a thousand kilometers per second. The emission was caused by a rising "bubble" of electrons behaving as if they were at a temperature of about 300 million degrees. (Copyright of CSIRO Division of Radiophysics, Australia)

a map of radio emission versus height in the corona, a quantity we can visualize more easily than radio frequency.

The radio emissions in figure 11.6 occur at the "plasma" frequency. This is the natural frequency at which electrons oscillate when they have been displaced. Their electric charge tends to make them return to their original position much as gravity forces a pendulum to return to its vertical position. The plasma frequency decreases as the gas density in the corona decreases. Emissions at 300 megahertz occur near the base of the corona; emissions at 30 megahertz occur where the density is 100 times lower, roughly one solar radius above the base. The dynamic spectrum observed from 300 to 30 megahertz covers the same portions of the corona as are visible during eclipse. The data in figure 11.6 range all the way from 2,000 to 7 megahertz, and from unusually dense parts of the coronal base right out to lofty coronal streamers. Spacecraft observe at still lower frequencies, generated in interplanetary space. Near the Earth, the radio emission occurs at roughly 30 kilohertz.

The various radio emissions all helped to open up a new vista on flares. They all indicated flare activity in the corona, which was unobservable in hydrogen alpha and much more energetic than expected. They demonstrated that flares have a "nonthermal" component that is totally unobservable in hydrogen alpha. The phenomena of figures 11.5 and 11.6 were regarded as

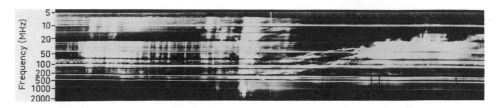

Figure 11.6. Dynamic radio spectrum observed in Australia in 1963. The recorded emissions (white areas) span a time of 25 minutes. The vertical axis can be interpreted as the height of emission in the corona. The steep streaks represent sources of emission sweeping upward through the corona at about one-third the speed of light. They are caused by beams of fast electrons, released dozens of times in this short interval, almost like rifle fire. The more gradually ascending pattern starting near the middle of this time interval represents a shock wave, released by a flare and traveling upward at a speed of roughly a thousand kilometers per second. (The vertical axis is labeled with the observed radio frequency in megahertz, where 1 megahertz = 1 million cycles per second, ranging from dense portions at the coronal base emitting at 1,000 megahertz to coronal streamers emitting at 10 megahertz.) (Copyright of CSIRO Division of Radiophysics, Australia)

evidence for fast motions upward through the corona. They were the observational forerunners of the transients such as those in plate 5 and figures 9.2 and 9.3.

Fast Electrons

Figure 11.6 shows many nearly vertical streaks traversing the corona in a few seconds. Whatever causes the emission must traverse the corona rapidly, in merely a few seconds. The moving emitters are known to be electrons, traveling as an electron beam. The typical speed of a beam is about a third of the speed of light, some 20 times faster than typical thermal speeds of coronal electrons. The beams occur many times during the 25-minute interval of figure 11.6, but they occur especially frequently, much like rifle fire, at the time of the flare, which is in the middle of the time interval of figure 11.6.

Electron beams have been of great practical interest in laboratories and have been investigated intensely for the last two decades. Machines for research on the future peaceful use of nuclear fusion energy have frequently suffered repeated discharges that create beams of electrons much like those in figure 11.6. Our laboratory and theoretical understanding of electron beams can now be tested in space. For example, many of the coronal electron beams can be measured when they reach the Earth, along with the radiation they emit. Apparently the theory for the radiation is still inadequate, although it is probably on the right track.

Radio events such as "Westward Ho!" (figure 11.5) provide still more evidence for fast electrons. In this case, the fast electrons are not in a beam but confined within a "bubble" of coronal gas rising upward through the corona after a flare.

Shocks

Figure 11.6 also shows a pattern of emission that rises much more gradually. It travels upward at a speed of a few hundred, perhaps a thousand kilometers per second. This is a reasonable speed for a shock wave launched by a flare. The explosive nature of the flare is likely to cause a shock wave, much as the explosion of a nuclear bomb does.

Shocks such as those indicated by figure 11.6 have been detected in many other ways. In the 1960s, G. Moreton detected major solar prominences shaken after a flare even though they were one solar radius away from the flare. Apparently the prominences are shaken by the impact of shock waves emanating from the flare. Most shocks from flares travel outward, including the shock observable in figure 11.6. Shocks are also associated with flareless transients, which are presumably caused primarily by magnetic fields ready to expand into space. Indeed, all shocks may gain additional energy from such magnetic fields. Many shocks observed reaching Earth appear to have energies several times more than the flare could have provided.

Figures 11.5 and 11.6 are from the 1960s. Since then, electronics technology has improved more than tenfold per decade. The information on solar radio emission has increased dramatically and, in the opinion of many, overwhelmingly. European astronomers, in particular, have worked on improving time resolution and obtaining high sensitivity to small changes in the radio emissions. A time resolution of 0.01 second is now available when needed. The improvement in time resolution was limited at least in part by the ability to evaluate the data. There is no point in acquiring such a vast quantity of data that one cannot even check whether they are interesting. Fortunately, these technical advances proceeded apace with another electronic development, the computer.

The various electronic improvements allowed scientists to detect many other kinds of radio bursts. Some have merely been classed by number, some have been given imaginative names like zebra or tadpole, as well as more prosaic names like blip or spike. They have given theoreticians much food for thought. Ultimately, they may provide valuable tests for our understanding of the more restricted plasma phenomena in terrestrial laboratories and in the Earth's magnetosphere.

The Coronal Flare

By the early 1960s it became clear that not only the hydrogen alpha but also the radio emissions were only a sideshow to the main event, an explosion in the corona. Space observations amply confirmed this expectation.

Early rocket flights detected copious X rays from flares. Figure 4.1, the X-ray picture taken just before the 1970 eclipse, happens to include a small X-ray flare, namely, the round structure slightly to the left of disk center. Plate 2 shows a more spectacular flare recorded by Skylab. Such X-ray emissions inform us that much of the energy radiated away by flares appears as X rays and that flares occur in gases of coronal temperatures. However, such X-ray pictures are a poor method of determining the actual temperature of the flare.

One of the Orbiting Solar Observatories provided a spectrum of a flare that included radiation from iron ions in which only a single electron remains (FeXXVI, in the notation of chapter 4). This high degree of ionization implies a flare temperature of about 20 million degrees. Suggestions of such temperatures in the early 1960s had met with great skepticism. But the new observations were convincing. Flare temperatures do reach 20 million degrees. Accordingly, Skylab and especially the Solar Maximum Mission carried equipment to measure flares at many wavelengths in the ultraviolet and X-ray ranges, appropriate to emission from gases well above 5 million degrees.

The normal Skylab X-ray observations were not taken often enough to follow the development of a flare. Skylab scientist-astronaut Edward G. Gibson, a solar astronomer, is credited with recognizing the beginning of a flare and adjusting the equipment so as to observe the development of the flare in considerable detail. When people indulge in the oft-repeated discussions on the relative value of manned versus unmanned space experiments, Gibson's flare observation is one example cited for the value of manned spacecraft. (However, the more recently launched unmanned Solar Maximum Mission is able to respond to a flare promptly.)

Flare Temperatures

Astronomers are as conservative as most physical scientists. They accepted flare temperatures above 10 million degrees only reluctantly and only after a spectrum had provided detailed evidence of such temperatures. But worse was yet to come.

The standard Skylab X-ray pictures are dominated by radiation in the wavelength range of 20 to 40 Å, usually called "soft" X rays. Skylab also carried an experiment to measure "hard" X rays, in the range of 2 to 10 Å. The shorter the wavelength of the X rays, the more energetic must be the electrons that emit the X rays. The "ordinary" flare temperatures up to 20 million degrees involve too few fast electrons to yield any measurable flux of hard X rays. Yet hard X rays are clearly observed. They last a much shorter time than soft X rays and they tend to be strongest near the impulsive phase.

What is the nature of the electrons emitting hard X rays? Are they simply very hot? That is, is the temperature of the gases in flares 100 million degrees or more? Alternatively, are there beams or clouds of near-relativistic electrons, as are already indicated by radio bursts? Electrons with speeds between one-third and two-thirds the speed of light would have ample energy to create most hard X rays upon collision with thermal ions. This question could not be resolved by Skylab data. The Solar Maximum Mission was designed in large measure with this goal in mind.

Hot Stuff

Do the hard X rays indicate high temperatures or fast electrons? The answer is sometimes one, sometimes the other.

A rather extreme example is the flare of June 7, 1980. The Hard X-ray Burst Spectrometer aboard the Solar Maximum Mission recorded the data shown in figure 11.7, with a time resolution of 0.128 seconds. The flare intensity was unusual in its oscillatory time dependence. The main maximum of the burst was resolved into seven peaks. After the underlying longer-term emission was removed, the peaks had a recurrence period of about 8 seconds (see inset in figure 11.7). When plasma physicists are shown these spikes, they promptly relate the spikes to what they call "saw-tooth" X rays in fusion machines. Fusion machines are built to create 100-million-degree gases, but the saw-tooth X rays indicate failures due to short circuits caused by repeatedly generating beams of electrons. Therefore, laboratory tests suggest that the hard flare X rays are related to electron beams. How sound is the analogy?

The flare of June 7, 1980, produced X rays at unusually short wavelengths, at least down to 0.03 Å. The most surprising feature of the flare is apparent when the intensity of the X rays in various energy bands is plotted separately at the time of one of the maxima and the time of the subsequent minimum of the intensity oscillation (main part of figure 11.7). The spectrum at the time of the intensity peak has the shape expected from an exceedingly hot thermal plasma, at 1.4 billion degrees!

The duration of the billion-degree "temperature" for only a few seconds causes a considerable theoretical problem. Normally, electrons and protons become "thermalized" by exchanging energy and momentum through many collisions. However, no time

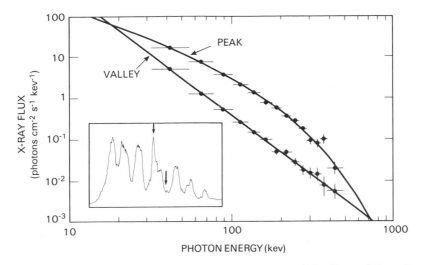

Figure 11.7. Billion-degree flares? Observations of the flare of June 7, 1980, with the Hard X-ray Burst Spectrometer aboard the Solar Maximum Mission. The flare included X rays of unusually short wavelength, which would not be emitted by a gas at "merely" 20 million degrees. Inset: The run of X-ray intensity versus time for about one minute shows seven peaks recurring every eight seconds. Such regularity is reminiscent of discharges in fusion experiments. Main diagram: Curves show the number of X rays versus the energy delivered when the X rays hit the detector, at two instants of time identified in the inset. At the time of the intensity peak, the spectrum corresponds to a "temperature" of 1.4 billion degrees! Merely four seconds later, the spectrum is already quite different, specifically no longer "thermal" in shape. (A definition of the energy unit Kev = kilo-electron-volts appears in table 2, chapter 13. X-ray wavelength = 12 Å divided by X-ray energy in Kev.) (Courtesy A. Kiplinger, *Astrophysical Journal*, and National Aeronautics and Space Administration)

is available for such collisions. Plasma processes have been invoked as a method of thermalizing the electrons, but pending convincing details, skeptics will doubt that figure 11.7 really represents thermal electrons near a billion degrees. The skeptics prefer to invoke relatively few near-relativistic electrons moving through flare gases with more acceptable temperatures. But they must admit that the strikingly thermal shape of the spectrum is merely an accident. Perhaps the argument can be settled after more flares are observed during the next solar maximum.

Dynamic Flares

Skylab could only hint that flares are very complex. The Solar Maximum Mission incorporated instruments explicitly designed to trace the complexities of solar flares. The instruments were carefully selected and designed to encompass all kinds of radiation expected from a flare, because each kind provides at least one important piece of the flare puzzle. Two instruments measure gamma rays and very hard X rays. No attempt was made to measure the position of their sources. Four instruments can obtain images of X rays and ultraviolet radiation; that is, they produce spectroheliograms at selected wavelengths. Then there is the white-light coronagraph, which detected transients such as the one in figure 9.3, and the irradiance monitor, which measures the entire solar energy reaching Earth (See chapter 14).

Every instrument was the result of design compromises among various degrees of angular resolution, wavelength resolution, time resolution, and expectations about what the flares would actually produce. Each instrument was designed around its special capability, with lower priority given to the search for other information. Yet the data from any one instrument can be interpreted meaningfully only if much other information is available. How much information can one instrument be designed to ignore, either because another instrument provides it or because the data can be obtained from ground-based optical or radio observations? For instance, gamma rays are observed with no spatial resolution at all. Can the source of the gamma rays be identified by simultaneous events appearing in pictures taken with high time and spatial resolution, either in soft X rays, or in the ultraviolet, or in the hydrogen alpha, or even with a radio telescope? In this case, the answer appears to be yes.

The interdependence of the instruments in space and on the ground led to a major international program designed to foster simultaneous observations from several optical and radio telescopes in coordination with the Solar Maximum Mission.

The entire set of flare data—optical, radio, and from space—must be interpreted consistently by each participant. This is a major, time-consuming effort, not only because investigators from many countries must learn how each other's instruments behave, but also because flares are vastly more dynamic and diverse than was suspected even after Skylab. The data must be analyzed almost second by second to identify the rapid ebb and flow of energies. Every interpretation must take into account the many simultaneous spatial components of the flare, each of which may involve different motions, gas densities, magnetic fields, and temperatures.

Although the total energy released by flares is small by solar standards, flares are of great interest because of their analogy to fusion machines, their radiations in the ultraviolet and X-ray regions that affect the Earth's atmosphere, and occasionally the release of additional coronal energy resulting in interplanetary shocks and further terrestrial consequences.

The Solar Maximum Mission has increased scientists' confidence that we can indeed understand flares. The detailed data have justified the investment in the mission and in its repair. Most of the Mission's experiments are still working and may yield new insights if they survive until the next solar maximum. Unfortunately, if the new solar cycle is as active as the previous one (see chapter 6), the Solar Maximum Mission will reenter the Earth's atmosphere and be destroyed as early as 1989. The Solar Maximum Mission has not generated much public enthusiasm about solar science, perhaps because most of its results appear as detailed graphs and there are not as many spectacular pictures as there were from Skylab. Of course, the detailed graphs and arguments are essential for accurate science. Nevertheless, the solar community has been somewhat remiss in its efforts to translate the results into accounts that the public can understand. Perhaps this omission is making it more difficult to obtain approval for the next major solar space experiment (see chapter 17).

Chapter 12

A Stressful Situation

Model of a Flare

The enormous amount of detailed flare information provided by the Solar Maximum Mission could not be interpreted without some type of model of a flare. No flare fits any one model. But most flares have some features that fit. A model is useful even if a flare does not fit it, because features that do not fit can be discussed in terms of how they differ from the model.

The favored model begins with a localized coronal explosion. One speaks of a "kernel" that almost instantly reaches temperatures of 100 million degrees and perhaps much more, depending on the interpretation of the hard X rays. The superheated gas expands promptly. The temperature decreases as the gas expands. We observe the resulting processes at various stages of expansion, depending on the instruments we use to monitor the flare. This model has survived for almost a decade.

How tiny is the kernel of the original explosion? The instrument imaging hard X rays cannot resolve its size. Instead, one estimates the size from the duration of the hard X rays. For instance, if the kernel expands at 2,000 kilometers per second, as expected for a 30-million-degree gas, and if the heated region doubles in size during 0.3 seconds, as suggested by the duration of the X-ray bursts, then the initial diameter of the kernel could have been no more than about 600 kilometers. Anything larger would require more time to expand and cool. These numbers are imprecise and vary from flare to flare, but the point is that the initial kernel of very hot gases is at most a few hundred kilometers across. On a solar scale, the explosion is remarkably tiny.

After the initial few seconds, the rate of expansion and evolution of the flare slows down. The Solar Maximum Mission can image the soft X rays from a flare when it has expanded to about 10,000 kilometers. At that point, the changes caused by expansion take place in roughly a minute. Thereafter, however, the geometry of magnetic fields becomes more important than time.

Magnetic Arches

Where is the flare kernel? Specifically, where is it situated relative to the underlying active region? Many times, the X-ray flare appears to occur in the corona over the line that divides oppositely directed magnetic fields at the photosphere. In terms of the coronal loops of figure 8.1, the flare occurs at the top of coronal magnetic loops. Figure 12.1 represents a plausible model. Although it is probably far too simple, it manages to incorporate a remarkable number of observations, both classical and space-based.

Chromospheric Explosion

The classical chromospheric hydrogen-alpha flare constitutes the "spills" from the coronal flare. When the flare kernel emits hard X rays, it contains many fast electrons, moving at a third the speed of light or faster. Up to 1 percent of these escape, to be replaced by electrons at more normal speeds. When the fast electrons reach the chromosphere, they collide and emit more hard X rays. Just such X rays are observed by the Solar Maximum Mission on either side of some flare kernels. During collision most of the electron energy is transferred to the local gas. The resulting sudden heating causes the classical hydrogen-alpha flare.

The sudden heating of the chromosphere by the fast electrons causes an explosion in the chromosphere. Much recent work has been concerned with detecting and measuring this chromospheric explosion. Are the observations consistent with the impact of fast electrons? Indeed, Richard Canfield of the University of Hawaii corroborates this picture by combining ground- and

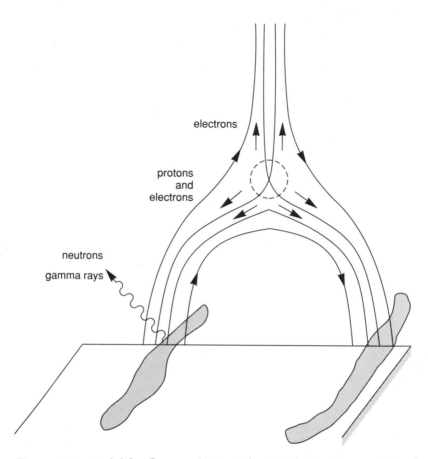

electrons

protons
and
electrons

neutrons

gamma rays

Figure 12.1. Model for flare explosion. The initial situation consists of a set of magnetic field lines connecting opposite magnetic polarities of an active region. They include a coronal loop much like the one in figure 8.1. The surrounding field lines reach high into an overlying coronal streamer. A geometrical "problem" is inevitable: Some field line must come to a point (circled). The flare kernel, the central explosion, occurs at such a point of localized electrical current. Electrons escape promptly, moving at a third of the speed of light (arrows). Those moving upward cause radio bursts. Those moving downward hit the chromosphere. The classical hydrogen-alpha flare is the footprint of the coronal flare. If the coronal flare occurs all along the top of a magnetic arcade, the hydrogen-alpha flare has the commonly observed shape of two ribbons.

space-based observations. From the site of the explosion, some gas moves downward at speeds of about 60 kilometers per second. It is observed in the hydrogen-alpha line. Some gas moves upward at speeds of about 300 kilometers per second. It is observed in an X-ray line of highly ionized calcium (CaXIX). Within the bounds of error, the two gases have equal and opposite momenta, as expected for sudden heating by a beam of electrons carrying much energy but negligible momentum.

The chromospheric explosion can also explain a long-standing puzzle, namely, the rather high density deduced for solar flares both in hydrogen alpha and in coronal line observations. These densities exceed the normal ones by one to two orders of magnitude. If the flare occurs in the corona, where does all this gas come from? Apparently it is expelled from the chromosphere early in the development of the flare. Only a tiny fraction of the chromosphere would have to be raised for this to occur.

The gas arriving in the corona becomes part of the coronal flare. It acts as a thermal sink and provides much inertia to further rapid expansion. It participates in the many phenomena that make up the gradual phase of the flare. Evidently, such additions of gas from the chromosphere may not be neglected. The chromospheric flare, after a decade of neglect, is assuming new respectability.

Pre-flare Activity

What is the cause of the flare? What supply of energy does it tap? What triggers the explosive release of energy? The model for the flare provides no answers. These questions have led to an almost separate line of research on pre-flare activity.

The only known source of energy that can be released quickly is the energy residing in magnetic fields and electrical currents. Somehow, electrical currents are dissipated into heat and motion. Certainly, currents dissipate rapidly if they are confined enough. Currents would have to be confined to widths of less than 1 kilometer to explain the release of transients and coronal heating by microflares (see chapter 9).

The problem is that an ordinary site of current dissipation is far too small. A volume of space 1 kilometer on a side cannot possibly contain the gas, the electrons, and the energy of a major flare. Somehow, the current dissipation triggered initially must release additional stresses, which are built up in a much larger volume before the flare begins. The majority of the flare phenomena arise when these stresses are relieved.

It is possible that the entire flare is a secondary phenomenon: Careful timing of transients that occur together with flares shows the transient starting before the flare. If so, the primary phenomenon is the release of the transient, which gives rise to motions that increase the stresses at any preexisting localized currents and cause the flare. If no such currents exist, the transient takes off without a flare.

Much of flare research and several workshops have concentrated on identifying the pre-flare stresses. Pre-flare activity has typically been studied during the last hour before a flare, but one goal is to recognize flare indicators as early as one day before a major flare. That would constitute a useful degree of flare prediction for practical purposes.

Much of the evolution of active regions has been described in terms of emerging flux tubes (see chapters 6 and 7). Flares clearly are most probable when magnetic flux is emerging rapidly. The Skylab discovery of X-ray bright points focused particular attention on newly emerging magnetic flux tubes.

The observations led to the "emerging flux model." Figure 12.2 shows the simple diagram usually associated with the emerging flux model, a diagram quite similar to figure 9.4, and also suggests the possible complications in real emerging magnetic fields.

There is some tendency for very energetic flares to recur with a period of about 152 days. If this periodicity is real, the fields emerging to cause any one flare are likely to be anchored deep inside the Sun, perhaps in the region that gives rise to the solar dynamo.

Vector Magnetograph

Further observational support for magnetic stress comes from the "vector" magnetograph. It measures not only the component

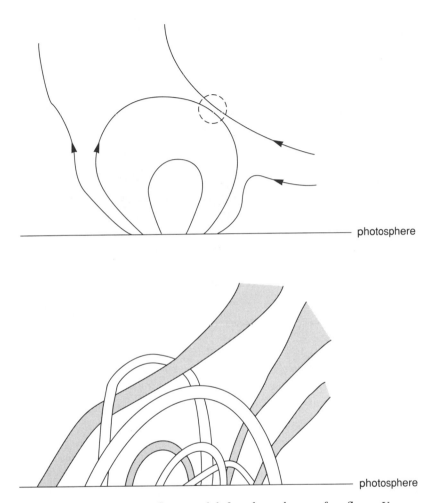

Figure 12.2. Emerging flux model for the release of a flare. Upper sketch: Snapshot of magnetic flux emerging from the photosphere. The new flux forces the approach of oppositely directed field lines (circled), a site of localized currents. As more flux emerges and the entire pattern expands, the site of the localized currents rises upward with the pattern. It is likely to flare once it rises into the corona. Magnetic distortions and stress extend far beyond the emerging field lines. The flare represents a relief of those stresses. Lower sketch: A slightly more realistic view of numerous emerging flux loops. Such a geometry is easily visualized but totally intractable mathematically.

of the field along the line of sight, as has been done for decades, but also the component at right angles to the line of sight. The newly measured component turns out to be a much better indicator of the electrical currents that cause the pre-flare stresses. In order to map this component, the instrument must measure the polarization of light emitted at the wavelength of a selected spectral line. The first vector magnetograph was built in the 1970s, in time to complement observations from the Solar Maximum Mission.

Magnetic stress is caused by electrical currents. Strong electrical currents can be recognized by the associated twisted magnetic fields. The twisted structures in figure 5.11 strongly suggest electrical currents, but they cannot provide a measure of the currents. The vector magnetograph can. Figure 12.3 shows a magnetic map and the possible magnetic geometry leading to the flare of April 8, 1980. The extensive twisting of the magnetic field in this rather minor flare was quite unobservable before the advent of the vector magnetograph.

The magnetic twist is a direct measure of the electrical current: high twist, high current. Sites of high current are good candidates for flares. Clearly, routine measurements of currents will be valuable for flare prediction. A new generation of vector magnetographs is now under design.

Explosive Release

Why do the magnetic stresses build up until significant flare energies develop? Why are they not released frequently and in small doses, as may occur normally in the quiescent corona? The usual answer is that the buildup has to do with the gradual rising of the emerging magnetic flux shown in figure 12.2. During a few hours or days, the site of localized currents rises with the emerging magnetic flux into regions of lower gas density. As gas within the rising site drains back down, the density of the gas there also decreases. When the site reaches the transition region, the gas density decreases by a large factor. Consequently, the magnetic forces experience much less gas inertia. New explosive phenomena suddenly become possible that were too constrained by the earlier higher density. Thus a flare be-

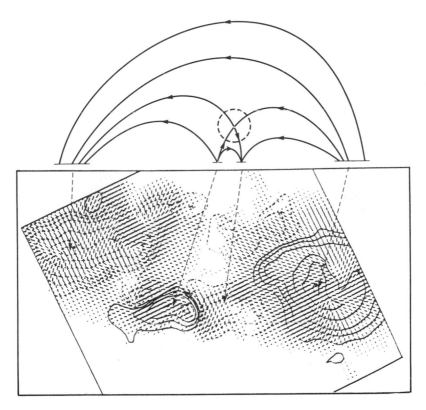

Figure 12.3. Magnetic geometry near the flare of April 8, 1980. The map represents the photosphere beneath the flare site as observed on April 6, two days before the flare. The short lines show the horizontal magnetic component (strength and direction) measured by the vector magnetograph at the NASA Marshall Space Flight Center. The horizontal fields change direction sharply in small areas, for instance, just below map center. Such magnetic twists are indicative of electrical currents and magnetic stress. The vertical components of the magnetic field are shown by contours, solid for up, dashed for down. The corresponding sketch of the coronal magnetic field lines includes two field lines almost making an X (circled), a likely site for current dissipation and flare release, quite similar to the sites in figures 12.1 and 12.2. (Courtesy R. Moore and *Astrophysical Journal*)

comes possible when the site of localized currents first enters the corona. The flare may develop as soon as conditions are ripe for the explosive phenomena or it may be triggered by, for instance, the extra stresses caused by a transient taking off.

The vector magnetograph indicates electrical currents only at the photosphere. It would be much more useful to have a direct measure of magnetic stress in the corona where the flare is likely to occur.

Very Large Array

Coronal magnetic fields can now be measured by the Very Large Array (see figure 12.4). Its 27 radio antennas act like a single radio telescope of very large diameter. The larger the effective diameter, the sharper the resulting radio picture of the Sun. Solar radio pictures can be nearly as sharp and detailed as the hydrogen spectroheliograms, at least when the radio emission is sufficiently intense.

The solar use of the Very Large Array became possible only after an administrative problem was settled. The 27 antennas were designed to track stars and galaxies. The Sun moves across our sky at a different rate. If the antennas were to track the Sun at the stellar rate, the maps made of the radio Sun would be smeared out. Tracking the Sun is technically not difficult. Nevertheless, to construct and install the appropriate electronics takes time and costs money. When money is in short supply, as it usually is, time and effort are needed to convince both nonsolar astronomers and administrators of the value of driving the antennas at the solar rate. The arguments quickly move from specific technical details to the general value of solar investigations. They become part of the "politics" of science. Mukul Kundu, of the University of Maryland, and his colleagues from other institutions were effective in having the solar tracking capability installed.

Coronal Magnetic Fields

The Very Large Array can map the Sun at several radio frequencies. It has been used mainly at relatively high frequencies, com-

Figure 12.4. Very Large Array near Socorro, New Mexico, consists of 27 radio antennas distributed along three 9-mile-long arms of a "Y." Together they can map intense radio flares with the same resolution as that of ground-based photographs. (National Radio Astronomy Observatory/AUI)

monly 5,000 megahertz. By comparison, the Culgoora radio-heliograph and the Clark Lake Radio Observatory (figure 11.4 and plate 13) were restricted to less than 400 megahertz and the usual radio receivers of the 1960s ranged only up to about 2,000 megahertz (figure 11.6). The Very Large Array has one further advantage over Clark Lake: It can detect radiation in both senses of polarization, whereas the antennas at Clark Lake, being wound in only one spiral sense (plate 13), could detect only one sense of polarization.

During flares, radio emission is so strong that a map of the radio emission can be made at the Very Large Array in merely 10 seconds. Such a "snapshot" appears in figure 12.5. One part displays a map of radio intensity. The second part is a map of the radio polarization.

How can we interpret such a radio map? The polarization

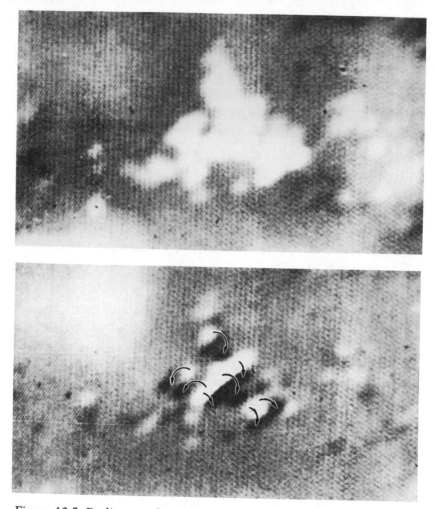

Figure 12.5. Radio snapshot of a flaring area taken during a 10-second interval on June 25, 1980, with the Very Large Array tuned to the radio frequency of 5,000 megahertz. The two maps inform us about phenomena at that height, in the low corona, where the magnetic field strength is about 550 gauss. The smallest recognizable structures are no larger than about 2,000 kilometers. Upper map: Intensity. Higher brightness indicates more numerous near-relativistic radiating elections. Lower map: Polarization. Bright indicates a magnetic field oriented toward Earth, dark away from Earth. Any pair of dark and bright areas of the polarization map indicate a magnetic flux tube protruding through the 550-gauss surface. The sketch of some possible flux tubes is not unique because any one white area in the polarization map might contribute flux to two or more dark areas. Nevertheless, magnetic stress must be high. No wonder the magnetic field caused a flare! (Courtesy M. Kundu and *Astronomy and Astrophysics*)

tells us that the radiation was emitted from a magnetized region. The details of the radiation process tell us just how magnetized the region must be. One finds that the polarized 5,000-megahertz radiation must arise from places in the low corona where the magnetic field strength is 550 gauss. Such strong coronal magnetic fields occur only in the vicinity of sunspots, where the fields in the underlying photosphere are typically 1,000 to 2,000 gauss.

The radio map is a form of spectroheliogram that tells us what happens on a surface at a certain height in the corona, in this case a surface at a height where the field strength is 550 gauss. In particular, the polarization map tells us where magnetic fields come up through that surface, and where they head back down.

How can we mentally connect the field lines passing the white and dark regions? The second part of figure 12.5 represents one attempt to do so. It indicates a central arcade of field lines and some subsidiary ones. The magnetic connections could be validly drawn quite differently. Nevertheless, the magnetic pattern clearly is not simple. Twists in the magnetic fields, though not drawn into figure 12.5, are almost inevitable. Stresses abound. No wonder the region flared!

The physical height of the 550-gauss surface is unknown, except that it is in the low corona. It is not yet possible to connect the radio magnetic map to the magnetic map at the photosphere, partly because neither map is sufficiently accurate, partly because the relative positions of the maps are not known well enough, but mainly because the radio map changes drastically in the course of a minute, whereas very little happens at the photosphere. In fact, a comparison of consecutive radio snapshots suggests the emergence of magnetic flux tubes through the 550-gauss surface. This is quite consistent with the emerging-flux model of flares, but again the conclusion depends to an unknown degree on the patterns visible to the eye of the beholder.

The observations are not yet sufficient for a detailed interpretation. One would like simultaneous maps at many more radio frequencies so that the flux tubes can be traced through surfaces at various heights. A "frequency-agile" receiver has recently been constructed for just such a purpose. It can map the flare at

80 radio frequencies nearly simultaneously. In effect, it will map the magnetic pattern at 80 heights in the corona. It will be used on a three-antenna network at the Owens Valley Radio Observatory in California, operated by the California Institute of Technology. The next solar maximum should provide the needed flares.

Flare Buildup

Figure 12.5 could be produced photographically because the radiation during this flare was quite intense. A large range in intensity could be measured, even with a "snapshot" exposed for only 10 seconds. The "buildup" of the magnetic pattern before the flare is more difficult to observe because the radiation is weaker before a flare. Typical pre-flare exposure time at the Very Large Array has been 15 minues. During this time, new polarized regions may appear on the map, which is in line with expectations from the emerging-flux model. Surprisingly, some features in the map reverse their polarization. Has the direction of the magnetic field flipped? Perhaps the radiating electrons have simply entered and "illuminated" a different flux tube. In any case, the radio observations clearly detect significant changes in the corona well before the impulsive phase of the flare.

There is no question about the pre-flare state: It is highly stressed and is changing significantly. The flare relieves the stress in a large volume of the corona.

Flare Warnings?

When will a flare occur? We would like to predict large flares well enough to be ready for their terrrestrial consequences (see chapter 14). Pre-flare radio heating and changes in radio polarization maps may provide the best opportunity to forewarn the public and researchers of flares. To be useful, however, such changes would have to be recognized and monitored in real time.

The real-time display of a map from the Very Large Array is still a very difficult undertaking. Even a single attempt to construct a radio map takes considerable computing. The data must be passed repeatedly through the computer until one can be fairly certain that small features on a radio snapshot are real. Until recently, such "clean" maps of the Sun took many days to produce. A delay of days will not do for a flare warning. Perhaps the analysis may be speeded up as experience in the production of solar radio maps increases and as ever faster computers are employed. However, a more fundamental problem will remain.

The major radio telescopes needed for real-time solar mapping are also used for many other kinds of observations. They are not available for the kind of routine solar surveying that flare warnings would require. The National Oceanic and Atmospheric Administration's Space Environment Laboratory plans on using satellite X-ray cameras routinely to observe the Sun and transmit the data promptly. So far, however, we know of no clear X-ray signal warning of a forthcoming flare. Useful flare warnings are still at least a decade away.

Chapter 13

Cosmic Rays and Fermi

Energy Ranges

Terrestrial speeds are much slower than the speed of light: A supersonic jet fighter plane may fly at a millionth the speed of light; even the Space Shuttle orbits the Earth only 20 times faster than the jet. Speeds tend to be higher in the solar corona: Transients rise at about a thousandth of the speed of light; radio-emitting electrons move at about a third of the speed of light.

Solar flares give rise to still higher speeds, albeit only of individual particles. Most of these move at close to the speed of light, so that their speeds are not very informative. Instead, it is convenient to think of particle energies, that is, the amount of work and damage they can do if they hit something. One speaks of relativistic particles, in the sense that Einstein's special theory of relativity is needed to describe their detailed behavior. Fortunately energy, their one most useful attribute, is a kind of quantity with which we are familiar on a daily basis.

Relativistic particles are extremely common in the Universe. Those that arrive here are generally called cosmic rays. That many more exist is deduced from the radiations they emit. Table 2 provides some perspective on observed energies. The range of observed energies spans 17 orders of magnitude. At the top of the list, the most energetic cosmic ray has the energy of a golf ball when it leaves the tee. Yet this energy resides in a single atomic nucleus, perhaps in a single proton!

Table 2. Particle Energies and Acceleration Sites

Units: 1 Tev (tera electron volt) = energy of a flying mosquito.
1 Tev   = 1000 Gev (giga electron volt).
1 Gev   = 1000 Mev (mega electron volt).
1 Mev   = 1000 Kev (kilo electron volt).

Energies	*Particles and Acceleration Sites*
10^8 Tev:	Highest observed energy of cosmic rays Acceleration at shocks surrounding galaxies
10^3 Tev:	Gamma rays from the stellar object Cygnus X3 Collisions by protons at several times this energy
10^2 Tev:	Energy of gamma rays from electrons in Crab Nebula Acceleration by plasma waves in supernova remnant
20 Tev:	Protons in planned Superconducting Supercollider Electric acceleration
10 Gev:	Average energy of galactic cosmic-ray protons Acceleration at shocks surrounding supernova remnants
1 Gev:	Energy of solar cosmic rays, mostly protons Acceleration at shock departing from flare site
100 Mev:	"Low-energy" galactic and solar cosmic rays Like Gev cosmic rays but more abundant
10 Mev:	Energy of observable cosmic neutrinos Nuclear reactions in solar and supernova interior
1 Mev:	Gamma rays from solar flares and stellar bursters Collisions by protons at several times this energy
100 Kev:	Radio-emitting electrons during solar flares Acceleration by plasma waves at flare
1 Kev:	Energy of interplanetary ions Carbon, oxygen, etc. with flare "thermal" energies
0.12 Kev:	Electrons in short circuits of household appliances Acceleration across 120 volts

Note: Particles are "relativistic" if their kinetic energy exceeds Einstein's famous quantity mc^2, which is 0.9 Gev for protons and 0.5 Mev for electrons; 10^8 stands for 10 to the 8th power, etc.

Accelerators, Cosmic and Terrestrial

How can individual particles receive so much energy? Although velocity is not a good measure of energy, one still asks how these particles are accelerated. Some stars and galaxies apparently provide powerful sites of acceleration. Table 2 identifies some possible sites.

Particles from solar flares have nowhere near the highest energies observed. However, they are of special interest because their process of acceleration can be observed in far greater detail than any comparable phenomenon at even the nearest stars. Solar flares are at least five orders of magnitude closer to Earth than any other stellar accelerators. They are nine orders of magnitude closer than any galactic accelerators. Therefore, despite their physical inaccessibility, solar flares may be taken as the most scientifically accessible site of cosmic particle acceleration.

Can we not explore particle acceleration at a much closer location in the laboratory? Indeed, the planned Superconducting Supercollider is at the forefront of such experiments. Its particle energies exceed those of solar flares (table 2). However, that machine is based on extreme order and precision. Nature is different. It is highly disordered. A more relevant and much larger "laboratory" for particle acceleration is now available within the Earth's magnetosphere and especially at its bow shock (see figure 10.5.). Thanks to numerous space experiments, particle acceleration at the bow shock and the geomagnetic tail is now understood well enough to generate confidence in the physical concepts. Computer simulations have played a major role in testing this understanding. However, particles from solar flares reach energies some three orders of magnitude greater than those accelerated in or near the magnetosphere. Thus, the solar flares may well provide the next "laboratory" for particle acceleration.

Historically, the most energetic particles from solar flares were observed first. They are detected more easily and they reach ground-based detectors more readily. Thus the first solar cosmic rays to be recorded from the Sun, during the 1940s and 1950s, had typical energies in the Gev range (table 2). Particularly important for early solar detection was a worldwide network of neutron monitors. The neutrons are the products of high-energy protons striking the Earth's atmosphere. Since protons from solar flares were anticipated, the neutron monitors were equipped with warning signals in case there was a sudden rise in counting rate. One sudden rise occurred in 1949. My family and I were having dinner with John Simpson and his wife at the time. Simpson, at the University of Chicago, had been one of the prime movers in setting up the neutron monitor network. During that dinner a phone call alerted him that all the counters were showing an enormous event. We did not see him again that night. The flare and its cosmic rays were discussed for years after.

The space probes of the 1960s extended the observable energy range down to the level of 100 Mev. Such events were much more common. Moreover, space detectors could measure directly the properties of the solar particles, such as their energies, charges, and masses. During the 1970s, space physicists such as G. Gloeckler of the University of Maryland constructed space-based instruments to record and identify particles in the Kev range, particles that move at speeds not much faster than thermal speeds in the solar wind. About a dozen planetary space probes now carry these instruments. They transmit information about flare particles and flare-caused shocks that have traveled outward from the Sun as far as the orbit of Neptune.

Unfortunately, the time and location of the acceleration at the Sun cannot be deduced accurately from particles reaching the Earth. These particles have experienced rather circuitous motions in the magnetic field of the solar wind (see chapter 10), and they may easily have taken two or three times as long to reach Earth as they would have needed on a straight path. Consequently, much essential information on particle acceleration at flares has been derived from the radiation emitted by the particles at the Sun.

Electron Acceleration: How Fast?

An important observational question has been: How long does it take for particles to reach, say, 100 Kev or 100 Mev? The more rapid the observed acceleration, the more restricted the explanation of the acceleration. Different answers have been derived from detectors on different spacecraft, sensitive to different particles at different energies that behave somewhat differently. The many data are being sorted in relation to two major categories of particles: (1) Electrons ranging in energy up to about 100

Kev and emitting radio and x-radiation (see chapter 11), and (2) protons in the energy range of a few Mev and up.

Over a period of years, a consensus has developed that electrons are accelerated to some 100 Kev in roughly 0.1 second, as indicated, for example, by the radio bursts (see figure 11.6) caused by electrons at about a third the speed of light (energy about 30 Kev). The beams appear to be sent on their way within an interval no longer than 0.1 second. A more direct indication of rapid acceleration is provided by bursts of hard X rays observed with the Solar Maximum Mission. The observations of figure 11.7 imply that many electrons reached energies of 400 Kev or more in less than 1 second. Even faster changes have been observed for other flares.

The radio and X-ray observations refer to different electrons: Those emitting radio waves move up, those emitting X rays stay in place or move down. Probably, however, the two groups are accelerated together in the low corona and their observations refer to the same process of acceleration.

Efficiency

Flare X-ray observations indicate a very large number of electrons, possibly carrying a substantial fraction of all the energy released in a flare. A debate has arisen as to how efficient the acceleration must be to yield that much energy in fast electrons.

One view is based on the few X-ray burst spectra indicative of a thermal gas with a temperature of up to about a billion degrees; such bursts were observed during periods of the flare of June 7, 1980 (see figure 11.7). If the spectrum of the hard X-ray bursts really implies that all the electrons in the source move with a thermal energy distribution, then the acceleration efficiency is 100 percent. There are only some vague notions on how such a state might come about. This "thermal" interpretation of the X rays generates much skepticism.

The opposing view is that some modest fraction of all the electrons are accelerated to the energies indicated by the hard X rays. Thus most of the flare gas is at more "ordinary" temperatures, some 10 million degrees, and emits primarily soft X rays.

In that case, the theorist must make detailed models to explain how and where the fast electrons emit hard X rays, taking into account all the observations on timing and location of the observed X rays and of the thermal-appearing spectrum. Neither theory appears to work for all flares.

Both views agree that as much as 10 percent of all electrons in the flare may reach the energies at which they emit hard X rays. Classically, any acceleration efficiency exceeding 1 percent was considered suspect. But opinion has changed with time. One now accepts that a very substantial fraction of the energy in a supernova remnant is used for the acceleration of cosmic rays. Therefore, the suggestion that solar flares achieve 10 percent efficiency is no longer dismissed. On the contrary, theorists now would like to use solar flare observations to support theories for efficient particle acceleration elsewhere in the universe.

One might expect electrons to be accelerated in the most central, most energetic parts of the flare. However, the site at which the flare is triggered, that is, the site of current dissipation, is restricted to a volume of space so small that it cannot possibly contain as many electrons as are accelerated in a flare. Acceleration must occur in a volume of space much larger than the presumed "trigger" for the flare. During the acceleration process, particles are probably picked up that have already been accelerated by more normally occurring processes, such as frequent microflares (see chapter 9).

Plasma Acceleration: Surf Riding

The burden of explaining particle acceleration now shifts to the theorist. There are two basic modes of acceleration, one most appropriate to the electrons with energies up to about 100 Kev, the other appropriate to the more energetic protons. These are known as plasma acceleration and Fermi acceleration, respectively.

Very intense electrical currents are expected in solar flares. Sufficiently intense currents drive localized oscillations in the electrons; these oscillations travel as waves called plasma waves. Such waves are known to occur with strong laboratory currents.

The radio emissions recorded in figure 11.7 are caused by plasma waves, which, in turn, are caused by electron beams and shock waves traveling away from the flare. Thus plasma waves too are probably created within the flares.

Once plasma waves are established, a few individual electrons can ride with the waves, much like surf riders. They gain energy as they are accelerated to the speed of the waves. Some simple estimates make this form of acceleration a very attractive explanation.

What electrons can be accelerated? Experience with plasma physics and fusion machines indicates that the fastest 0.1 percent of the electrons are picked up and further accelerated by the waves. Even before any acceleration, these electrons carry roughly 1 percent of the thermal energy in the gas. If such pickup occurs in a pre-flare gas at 5 million degrees, the selected electrons already have an energy of at least 0.4 Kev.

As the selected electrons speed up, their energies are increased 10- to 100-fold. Picked up at 0.4 Kev, they reach between 4 and 40 Kev. This energy range matches the energies deduced from most of the X-ray bursts. Admittedly, the 400 Kev needed for the flare of June 7, 1980 (see figure 11.7), is somewhat extreme.

The energy efficiency is also reasonable. After acceleration, the total energy of the fast electrons constitutes between 10 and 100 percent of the energy in the remaining thermal gas. What really counts is the ratio of electron energy to the energy released from the relaxation of magnetic stresses (see chapter 12). If that energy is 10 times the thermal energy, then the fast electrons carry between 1 and 10 percent of the flare energy, which is similar to the value deduced from the observations.

Acceleration by plasma waves is easily fast enough to satisfy the observations, but it cannot account for the highest particle energies observed. A second mechanism is needed to reach these energies.

Fermi Acceleration: Ping-Pong

The first panel of figure 13.1 shows a pair of Ping-Pong paddles approaching each other slowly but inexorably, with a small

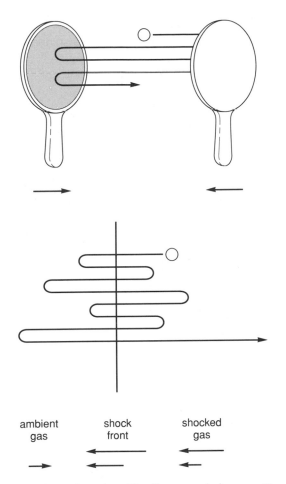

ambient   shock   shocked
gas   front   gas

Figure 13.1 Fermi acceleration. The first panel shows a Ping-Pong ball
accelerated by bouncing elastically off two slowly but inexorably con-
verging paddles. Arrows indicate paddle velocities. The second panel
shows the orbit of a proton between the gases surrounding a strong
shock. The upper set of arrows indicates the velocities we see: The
ambient gas is at rest; the shock front moves toward the ambient gas
more rapidly than the shocked gas. The lower set of arrows indicates
velocities in a frame of reference such that ambient and shocked gases
have equal and opposite velocities. The gases appear to converge like
the Ping-Pong paddles, accelerating the proton as long as the proton
keeps returning to the shock front. Shocks are the most likely cause of
acceleration of particles to the highest energies observed in flares.

Ping-Pong ball bouncing elastically and rapidly between the paddles. The ball always hits the paddles head-on and gains energy
at each bounce. When the separation of the paddles decreases by
10 percent, the energy of the ball increases by 10 percent. When
the paddles approach each other to merely 10 percent of their
original separation, the energy of the particle has increased 10-
fold. Of course, this cannot go on forever. Some assumption
breaks down. For instance, the paddles cease to accelerate the
ball when their separation is reduced to the diameter of the ball.

Enrico Fermi, then at the University of Chicago, suggested
that cosmic rays travel along interstellar magnetic field lines
and "bounce" off the stronger magnetic fields anchored in interstellar clouds, much as a ball bounces off the Ping-Pong paddles
of figure 13.1. He envisaged that acceleration would occur between two clouds that approach each other steadily ("first-order
Fermi mechanism") or between clouds that have random velocities, so that cosmic rays are sometimes decelerated and sometimes accelerated, with acceleration occuring slightly more frequently ("second-order Fermi mechanism"). It is now clear that
real interstellar clouds make leaky traps and cannot provide the
acceleration proposed for the cosmic rays. But Fermi acceleration survives as an absolutely essential concept.

Shock Acceleration

In the mid-1970s, it became clear that shocks would make good
first-order Fermi accelerators: The gases on the two sides of the
shock always approach each other. See the second panel of figure 13.1. If a particle is scattered by the gases on either side of
the shock front, then it has an opportunity to bounce back and
fourth across the shock many times. It is then accelerated just
like the ball between the two Ping-Pong paddles. Both theory
and interplanetary observations show that the scattering actually occurs. The scatterers are Alfvén waves (see figure 8.2), often
created by the particles they scatter.

Particle acceleration continues until some assumption breaks
down, just as in the case of the Ping-Pong paddles. Most probably, acceleration stops when the particle escapes from the vicin-

ity of the shock. Many bounces are needed for a particle to reach high energies. The higher the energy a particle reaches, the more often it must have bounced, the more chance it has to escape from the shock region, and the fewer such particles there are. In the simplest version of the theory, one obtains a power-law distribution in energy: The number of particles above a given energy is proportional to the energy to some power. A power law is an attractive explanation because observed cosmic rays actually have such an energy distribution, and so do the electrons deduced to exist from solar hard X rays. The theory for Fermi acceleration has been elaborated well beyond this simple version. The predicted energy distributions still work remarkably well, even when confronted with rather detailed particle observations spanning a large range of energies.

Until a few years ago, acceleration beyond about 1 Mev was thought to be relatively leisurely. A minute seemed reasonable, based on some long-lasting radio bursts and on the arrival times at Earth of some solar particles with apparently rather direct paths from the Sun to Earth. It seemed reasonable that Fermi acceleration could produce even the most energetic particles in about a minute.

Proton Acceleration: Crucial Observation

The notion of a relatively leisurely acceleration to higher energies has become quite untenable. The crucial observation, reported by E. Chupp of the University of New Hampshire, was based on the gamma-ray spectrometer aboard the Solar Maximum Mission. The gamma-ray data have a time resolution better than 2 seconds. During a flare of June 7, 1980, gamma rays in the range of 4 to 7 Mev arrived within two seconds or less of the impulsive burst of hard X rays (see figure 13.2). The gamma rays are emitted when protons carrying energies of about 10 Mev travel down from the flare and hit the chromosphere. The protons were apparently accelerated to 10 Mev in a mere 2 seconds. In another flare, neutrons observed at Earth indicate equally prompt acceleration of protons at the flare to at least 50 Mev (see figure 12.1).

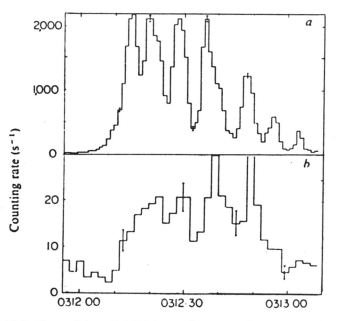

Figure 13.2. Flare of June 7, 1980 (same flare as in figure 11.7), observed by the Gamma-Ray Spectrometer aboard the Solar Maximum Mission. The top graph shows radiation intensity versus time much like the inset in figure 11.7, including the same seven peaks spread over one minute. The bottom graph shows the intensity of gamma radiation (gamma energies in the range 4.1 to 6.4 Mev), which must have been emitted by protons with energies of at least 10 Mev. Time resolution is about two seconds. The near simultaneity of the peaks in the two graphs indicates acceleration to the highest energies in merely five seconds or less. (Permission E. Chupp, adapted by permission from *Nature*, vol. 305, no. 291, figure 1, copyright 1983, Macmillan Magazines Limited)

In both flares, proton acceleration to the highest energies occurred in less time than previously expected by almost two orders of magnitude. This is yet another spectacular deduction and a significant technological feat of the Solar Maximum Mission!

How have theoreticians responded to such observations? Some believe that the observations indicate not the time needed for acceleration, but merely the time needed to release some particles that were accelerated and stored earlier. Indeed, interplanetary protons in the range 3 to 50 Mev have been observed that were apparently released from the Sun already three hours before a flare, and a flare might release more of such a supply of protons. But this explanation is not very satisfying. Most theoreticians simply deduce that acceleration occurs remarkably rapidly and theory must accommodate itself to the observations.

Is Fermi acceleration really fast enough? One has to envisage a flare core becoming heated, expanding, and creating a shock, which then accelerates particles, all within the observed limit of merely about one second. The best evidence supporting such a scenario comes from the geomagnetic bow shock (see figure 10.5), where theory and observations agree in considerable detail.

The period of the next solar maximum will provide further tests for our understanding of particle acceleration in flares. Barring surprises, by then the theory may be ready for cautious extrapolation to other cosmic sites of particle acceleration, particularly the acceleration of cosmic rays by supernova remnants. The solar flare will then provide a true cosmic laboratory.

Part IV

Perspectives

Chapter 14

Terrestrial Responses to Solar Changes

Dramatic Flare Effects

The white-light flare observed by Carrington in 1859 was followed, after a few hours, by northern lights over much of Europe and even Honolulu. Telegraph systems were troubled throughout the world. An exception was the telegraph system connecting Boston and Portland, Maine, which worked for two hours on naturally driven electrical currents, without any batteries at all.

Three large flares observed on August 2, 1972, were followed by even more dramatic effects. Technology had advanced in the intervening century and was more widely available to be disturbed. Two days after the flare, the current in a power line in Canada surged and a transformer exploded. Circuit breakers were triggered, interrupting power distribution to several states and provinces. Canadian telecommunications links were damaged. Navigation by high-frequency radio was interrupted. In the United States, AM radios tuned to distant radio stations lost the stations and received nothing but crackling noise. Other radios received stations that are normally far out of range.

Chain of Connections

Carrington noted that the flare was followed by unusual terrestrial phenomena. He could find no causal connection, but in fact his intuition was excellent. During the next century, so many more coincidences were noted that the idea of flares causing terrestrial phenomena was generally accepted. But proving the causal connection is difficult.

The frequency of the northern lights follows an 11-year cycle, just as the number of sunspots does. Is this evidence that the northern lights are caused by sunspots? The temptation is to answer yes. But the evidence is only circumstantial. Modern space observations show that in fact the connection is rather indirect: Sunspots are associated with flares; flares may cause interplanetary shock waves; the shock waves alter the Earth's magnetosphere (figure 10.5); the change in the magnetosphere may release fast electrons that cause the northern lights. This is a very long chain of connections.

Usually, such a long chain of connections should elicit skepticism at least until all the links can be plausibly established. Today, the links between flares and the Earth are indeed plausibly established, although many of the details are still poorly understood.

The flares are the most dramatic aspects of solar activity. There are numerous other apparent relations between solar changes and terrestrial consequences. They are frequently called "solar-terrestrial relations." All involve long chains of connections, some reasonably well established, some very tenuous indeed.

A major International Solar-Terrestrial Program is planned for the mid-1990s. Current plans call for six coordinated space probes designed to trace solar-terrestrial effects in considerable detail all the way from the Sun to the ground. U.S. participation will be integrated with its Global Geosciences Program.

Timing

The lack of a complete global explanation should not detract from our careful study of the manifestations. Flares have attracted the most attention not only because their terrestrial effects are dramatic but also because they are the most precisely observed changes on the Sun. By the 1950s, frequent routine hydrogen-alpha photographs permitted accurate timing of the maximum of solar flares. The time intervals between flare maximum and terrestrial effects became the first quantitative measurements to link these phenomena.

Prompt Effects: Communications and Solar Cosmic Rays

Some terrestrial effects occur at or soon after the time of flare maximum. The effects on radio communications are the most obvious. At radio frequencies of roughly 100 kilohertz, the AM band in North America, the signals from a transmitter may suddenly appear to come from a different direction. Reception may suddenly become very noisy because of the additional signals received from distant thunderstorms.

Normal long-distance transmission in the AM band makes use of a reflecting layer about 150 to 300 kilometers high in our atmosphere, a layer called the ionosphere (see figure 14.1). The reflection phenomenon was first recognized in 1901. It was tested experimentally in the 1920s and explored in detail during the International Geophysical Year, 1957–59. The gases in the ionosphere are normally ionized by solar radiation and the resulting electrons cause the radio reflection.

During a flare, excess ionization in our atmosphere causes the radio signals to be reflected at lower heights, often less that 100 kilometers. Space observations were needed to show that the excess ionization is caused mainly by ultraviolet emission of the flares.

Fast particles, mostly protons traveling almost at the speed of light, arrive minutes after flare maximum and may continue for several hours. By analogy with the cosmic rays arriving here at all times, the flare particles are usually called solar cosmic rays (table 2, chapter 13). When sufficiently fast particles collide with a space probe, they create many more new particles that are dangerous to living matter. The flare of August 7, 1972 (figures 1 and 11.1), created enough solar cosmic rays to be lethal to Apollo astronauts had an Apollo craft been in space then. Future astronauts on the way to the Moon or the planets may wish to avoid periods when great flares may produce many solar cosmic rays. At present, we are unable to predict such periods (see chapter 12).

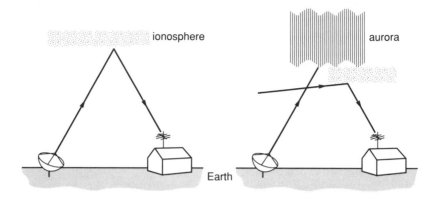

Figure 14.1. Radio transmission. Left panel: Normal transmission at radio frequencies below about 10 megahertz occurs by reflection off the ionosphere. Right panel: Ultraviolet from flares enhances ionization so that reflection occurs lower in the ionosphere. Later, ionization near northern lights causes the ionosphere simply to absorb the radio waves.

The magnetosphere prevents most particles from reaching equatorial regions near the Earth. Therefore, astronauts in the Space Shuttle and in future space stations with similar orbits are fairly safe from cosmic-ray damage. The safety of space stations in polar orbits or in the much higher geosynchronous orbits is still open to question.

Delayed Effects

The most important terrestrial flare effects start a few hours to three days after the flare. Most visible are the auroras, also known as the northern lights (aurora borealis) and the southern lights (aurora australis). Technically most important are the geomagnetic storms, which reflect changes in the Earth's magnetic field. They cause electrical currents in power lines and in oil pipelines and, indirectly, cause damage to sensitive electronic equipment on Earth-orbiting satellites.

The delayed effects start when the shock wave from the flare arrives at Earth. The normal solar wind takes between two and four days to arrive here. The shock travels through the wind and arrives somewhat more quickly, typically about a day after the flare but occasionally within a few hours. The arriving shock squeezes the sunward side of the magnetosphere (figure 10.5). The magnetosphere is then maintained in the squeezed state for many hours by the pressure of the enhanced solar wind arriving after the shock.

Aurora

The northern lights often appear as colorful "curtains" (see plate 14). What chain of connections can lead to such structures?

The place most strongly affected by the magnetospheric compression is the very front of the magnetosphere, facing the solar wind. In particular, the impact alters the electrical currents there. Some new electrical currents are forced to run along the magnetic field lines downward to the Earth. The currents are quite strong and may create sites of intense electrical sparking several hundred kilometers above the ground. Electrons are accelerated there much like the electrons causing sparks when a high-voltage power line breaks during a storm. These electrons then hit the atmosphere, collide with atoms and molecules, and excite them to radiation. Nitrogen may emit crimson light, oxygen greenish white light. Both colors are typical of the aurora. Since the original currents run on a thin sheet, the electrons continue on thin sheets, causing aurora in the form of thin curtains.

Just as the front of the magnetosphere responds to the shocked and irregular solar wind, so do the magnetic field lines carrying the electrical currents and electrons, and the auroral curtains appear to wave majestically in the atmosphere.

The geographical location of the aurora depends on which field lines at the Earth connect to the electrical currents at the geomagnetic boundary. Most auroras occur on an "auroral oval" centered on each magnetic pole. The northern magnetic pole is located in northern Canada. Therefore, northern lights are common in the populated parts of Canada but not in Germany, even though the two are as far north geographically. Northern lights in Germany are so rare that a red glow in the sky tends to be explained as a large fire beyond the next hill.

The stronger the impacting shock, the more it squeezes the magnetosphere, the closer to Earth is the geomagnetic boundary, the further from the magnetic pole are the feet of the active field lines, and the further from the magnetic pole one may observe the aurora. Only major flares can distort the magnetic field sufficiently for an aurora to be observed in Germany. An aurora was observed in Singapore, near the geographical equator, on September 25, 1909. It may have been southern rather than northern lights, since Singapore is closer to the southern magnetic pole.

Magnetic Storms

The compression of the geomagnetic field causes electrical currents to run throughout the magnetosphere, encircling the Earth. The currents, in turn, cause a change in the magnetic field at the Earth's surface. Both the direction and the strength of the magnetic field change, the latter frequently by 0.1 percent or even more. This is called a magnetic storm. The existence of magnetospheric electrical currents was deduced from ground-based magnetic measurements long before they could be verified by satellites.

In any electric device, a change in the electrical current flowing in one conductor alters currents flowing in neighboring conductors. Similarly, changes in the magnetospheric currents alter electrical currents flowing near the Earth's surface. When the magnetosphere is squeezed, following the impact of a shock from a flare, and currents are forced to flow throughout the magnetosphere, currents may also begin to flow through the solid Earth, through electrical power lines, and through oil pipelines. The new currents may arise in just a few minutes, they may last for hours, and they may disappear just as suddenly as they appeared.

Flare-induced currents in electrical power lines may cause voltage fluctuations of 50 percent and cause the normal current to arrive at the wrong phase of the alternating-current cycle. Safety equipment then shuts down the power transmission and causes blackouts. These electrical problems become more significant as power lines become longer and as civilization depends increasingly on an uninterrupted service of electricity. Unwanted currents may also damage the control equipment installed at intervals along the Alaskan oil pipeline, and they may contribute to the corrosion of the pipelines.

The strong changes in the geomagnetic field also redirect the fast electrons and protons that make up the Van Allen belts (see figure 10.5). The sudden arrival of such particles in normally "safe" places may damage satellites. The particles may degrade solar power cells or cause electric charges to accumulate on charge-sensitive measurement equipment. Radiation damage to electronic circuits becomes an ever greater threat as satellites depend increasingly on highly integrated and miniaturized circuits.

Magnetic Poles

Solar flare particles have direct access to the polar ionosphere. The polar ionosphere may be disrupted almost continuously for several days, from the time of flare maximum to the end of the geomagnetic storm. Also interrupted are most radio communications near the poles that depend on ionospheric reflection.

Historically, radio interruption was especially important in the polar regions because airplanes flying there lacked both visual navigation and reliable gyrocompasses. They tended to get lost without radio communications, especially during unpredictable but long intervals after solar flares. A solar astronomer, Robert C. McMath, earned military gratitude during World War II for alleviating this problem. He suggested that communications might be interrupted less frequently if the airplanes changed the radio frequency used for their communications.

Effects of Solar Activity

The geomagnetic effects of flares last at most a few days. The weather does not respond measurably to individual flares, to our knowledge. But the weather may well respond to more gradual changes on the Sun over months, decades, millennia, and even longer. Such changes may alter our climate, that is, the long-term global average of the weather. The changes in climate may be small, but they may be incomparably more important than the effects of individual flares. As the population on Earth grows and many more people move to marginally habitable areas, it will be more important than ever to measure and understand all the factors that influence our climate. Even small changes in climate due to changes on the Sun can take a heavy toll of human lives.

Solar effects on the Earth's climate have been a subject of debate for a long time. For instance, tree rings in some regions grow in an 11-year cycle. Are the trees affected by the 11-year sunspot cycle? Why then is the pattern not global? Currently, the skeptics are in the majority. They point out that the behavior of the stock market, outbreaks of Asiatic cholera, and the levels of the Nile and Thames also seem to be related to the solar cycle. Anyone seriously interested in solar-terrestrial relations would want to search for some reasonable explanation of the apparent relationships. The 1950s brought some initial steps in that direction.

Winter Storms

Low-pressure regions over the northern reaches of the Pacific Ocean have attracted attention because they tend to spawn major winter storm systems. Walter O. Roberts noted that these winter storms tend to be stronger if they are preceded by a geomagnetic storm. Geomagnetic storms tend to recur with a period of 27 days, complemented by isolated storms from solar flares. The Sun rotates with a period of 27 days. The coincidence in periods was taken as good evidence that the recurring geomagnetic storms are caused by solar activity. Therefore, Roberts concluded, the major winter storms are also influenced by solar activity. His arguments were taken lightly by his colleagues. Once again, the connection seemed too indirect.

New information emerged when space probes measured the solar wind, particularly the direction of the magnetic field carried past the Earth (see figure 10.5). The low-pressure areas in the northern Pacific tend to be weaker one or two days after the interplanetary field changes direction (see figure 14.2). In fact,

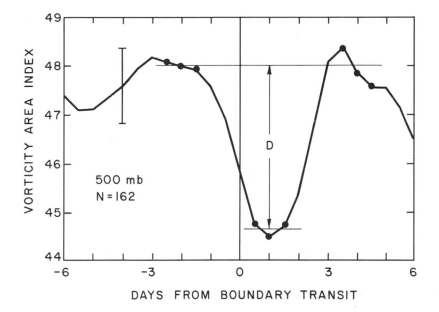

Figure 14.2. Solar effect on weather. The area in the northern hemisphere containing strong winter cyclonic storms is significantly decreased in the day or two after the interplanetary magnetic field changes direction. That direction, in turn, changes in response to solar magnetic fields rotating with the Sun. (Courtesy P. Scherrer; reprinted by permission from *Nature*, vol. 280, no. 845, figure 1, copyright 1979, Macmillan Magazines Limited)

the effect occurs preferentially for one of the two possible directional changes—that is, if the geomagnetic field and the field in the solar wind start almost parallel and the wind field is switched so that the two fields become nearly antiparallel.

The direction of the interplanetary field changes in response to the large-scale magnetic fields rotating with the Sun (see chapter 10). Here, finally, is a somewhat closer connection between solar and terrestrial phenomena!

Only a correlation has been discovered. How can distant interplanetary magnetic fields influence weather near the Earth's surface? The answer remains speculative: The reversal in the interplanetary field alters the electrical currents at the front of the magnetosphere; these alter the electrical currents throughout the magnetosphere; the latter currents alter the electrical voltage associated with the currents; finally, the voltage drop near the Earth influences the frequency of thunderstorms and thus the weather. Until such a causal relation is demonstrated piece by piece, most meteorologists are adopting a "wait and see" attitude. Perhaps they are justified, for once again an effect has been explained by a long chain of connections.

The correlation shown in figure 14.2 is based on data from the winters of 1963 to 1976. Several years passed before the correlation was taken seriously and some attempts were made to explain it. However, doubt about its interpretation has again increased because data for more recent years do not show this particular correlation. Meanwhile, more 11-year patterns have been discovered, related to the winter temperature in the stratosphere over the north pole and to the average latitude of winter storm tracks over the northern Atlantic Ocean. Strangely, the patterns can be observed only when stratospheric winds in the tropics are westward. It is a frustrating business.

Droughts

Roberts also emphasized a relation between droughts and sunspot numbers based on historical records in the American Midwest and West. Figure 14.3 is a partly updated version of his plot of the sunspot numbers with time. The relation of the 22-year sunspot cycle to the high-plains droughts is striking. If polar weather is related to solar magnetic fields, might the drought cycle correspond to the solar magnetic cycle?

The graph of figure 14.3 terminates in the early 1970s. Would the drought strike again? If so, it would strike in 1976. I lived in Boulder, Colorado, in early 1976 and knew nothing of the drought cycle at that time. But the drought was obvious even to a city dweller. The snowfall was minimal. Since snow provides water, agriculture was seriously hurt. So was the skiing industry. Perhaps it was fortunate that Denver had lost its proposal for the Winter Olympics of that year, because some of the skiing competitions would have been nearly impossible to arrange.

The thickness of tree rings can inform us of droughts further

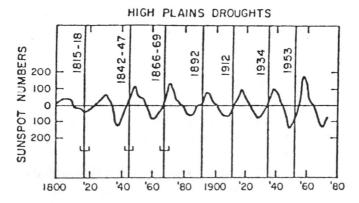

Figure 14.3. Droughts and the solar cycle. The years of droughts in the high plains of the United States are indicated on a plot of sunspot numbers. Spot numbers are plotted alternately up and down so that the 22-year cycle becomes evident. The droughts have occurred in phase with the solar 22-year cycle. The plot suggests a drought in 1976. It happened punctually. (Courtesy W. O. Roberts, National Center for Atmospheric Research)

back in time than can American historical records. Tree rings document a tree's response to the seasonal and annual variations in rainfall. Measurements of tree rings can be used to study climatic patterns during past decades and centuries, quite aside from any possible solar connection. They have been used to obtain data on high-plains droughts during centuries before any historical records were kept on these events. Such studies have shown that the droughts are closely linked to the 22-year sunspot cycle for a period of almost three centuries, as far back as we know the sunspot cycle, to about A.D. 1700. The impetus for seeking causal connections is now quite strong. Recent measurements of temperature changes in the atmosphere and even in the oceans seem to confirm that temperature varies with the solar cycle, but much careful study will be required before this can be established conclusively.

Carbon 14

The relation between solar activity and the climate can be carried even further back in time, if one uses a "proxy" for solar activity. The best proxy is an isotope of carbon, carbon 14. This radioactive variety of carbon is produced in the air by cosmic rays, at very nearly a constant rate. It is then gradually absorbed by living matter as part of its nutrient cycle. Once the matter dies, the carbon 14 decays slowly over several thousands of years. Because of its decay, less is found in older matter. "Carbon dating," a method of finding the age of matter whose age is not otherwise known, is based on a measurement of the carbon 14 remaining in the matter.

Of interest for the solar-climate connection are the slight changes in the carbon 14 production rate. To establish these changes, one needs plant matter of known ages. Tree rings have known ages. Judging from the tree rings of the last three centuries, slightly less carbon 14 is produced at solar maximum than at solar minimum. The relation is not surprising since fewer cosmic rays are measured near solar maximum.

If the same relation holds over many centuries—that is, if there is less carbon 14 during higher solar activity—then the carbon 14 in tree rings can inform us of long-term changes in solar activity for as far back as tree rings can go, about four thousand years.

The long-term relation between carbon 14 and solar activity is confirmed to some degree by the fact that periods of low carbon 14 are also periods with historical sightings of naked-eye sunspots, which are known to occur only at times of very high solar activity. It is also confirmed by the Maunder minimum.

Maunder Minimum

The period from A.D. 1645 to 1715 was a period with essentially no recorded solar activity, as noted by Maunder in a largely ignored article published in 1894 (see figure 6.2). Was this period merely a historical artifact? Was this a period with little interest in observing or recording celestial phenomena? Or was there actually a lack of solar activity? J. Eddy of the High Altitude Observatory searched historical manuscripts to demonstrate a lack of solar activity. His results, published in 1976, were impressive. What is perhaps most convincing, no northern lights were recorded for a 37-year interval during this period even though such unusual phenomena would surely have been noted. Moreover, the three solar eclipses during the Maunder minimum were all described in a manner consistent with very low activity.

The carbon 14 record shows a distinct maximum from A.D. 1640 to 1720, which is quite consistent with the lack of solar activity. Thus the carbon 14 record provides a proxy measurement of slow variations in the strength of the solar cycles.

Doubts

Can one use the carbon 14 records to show that solar changes cause climate changes? The answer is less positive than one might have hoped. At first, the Maunder minimum seemed clearly related to the "little ice age." No one doubts the existence of that cold period: It is evident in European records of failed crops. It is evident in American records of the American revolution, when General Washington crossed a frozen river that never freezes today. Glaciers are still retreating from the advanced positions they had reached over a century ago, near the end of the little ice age. But, contrary to initial expectations, the detailed timings of the little ice age match the solar Maunder minimum only poorly. Moreover, carbon 14 increases like that of the Maunder minimum are quite common, rather more common than "little ice ages" in historical records. Therefore, the Maunder minimum and the little ice age of that time are no longer considered strong evidence for solar influence on our climate.

Nevertheless, a long-term relation between carbon 14 and the climate seems to exist. Periods of high carbon 14, and presumably low solar activity, tend to be periods of cold climate, in accordance with the Maunder minimum. Perhaps Aristotle developed his concept of the perfect Sun only because he lived at a time of low activity, when no spots were visible to the naked eye. Periods of low carbon 14, and presumably high solar activity, tend to produce a hot climate. One period of low carbon 14, the warm period of about A.D. 1200, produced favorable conditions for the Vikings' sea voyage to America. It may also have caused the disappearance of the Anastasi Indian civilization from the southwestern United States.

Ancient Solar Cycles?

The carbon 14 records go back only a few thousand years. Ancient lake deposits in Australia, laid down about 600 million years ago, show fluctuations in the rate of deposit with time intervals of 9 to 12 years. These variations have been assigned to the influence of the solar cycle at the time of deposit. If that is correct, the influence of the solar cycle on climate 600 million years ago was much more important than it is today. However, the global atmosphere was different at that time, and one cannot yet estimate what climatic changes were needed to cause the observed variations in sedimentation, and certainly not what solar-cycle changes were needed to precipitate those climatic changes.

The "Solar Constant"

Meanwhile, the solar-terrestrial relation is being investigated in totally different realms. Is it possible that the luminosity of the Sun changes with time? We used to assume that it does not. The energy incident on an area of 1 square centimeter, facing the Sun at the top of the Earth's atmosphere and corrected for the annual changes in Sun–Earth distance, has always been called the "solar constant."

To Charles Greeley Abbot, for many years Secretary of the Smithsonian Institution, the answer to this question was maybe. He recognized that any changes in solar luminosity would affect weather, crops, and the future of civilization. He established solar observing stations worldwide in dry sunny sites with the goal of achieving better than 1 percent agreement among the stations and maintaining observations for better than 20 years.

Abbot was convinced that the Sun's luminosity fluctuated. He died in 1973 at the age of 101. Just a few years later, satellite evidence confirmed Abbot's opinion that the Sun's luminosity is changing, but it also showed that his measurements through the Earth's atmosphere were inadequate to measure the changes.

Space-based measures are difficult to make because one must frequently calibrate the observations with sufficient accuracy while the satellite is exposed to the vagaries of space. The first weather satellite carrying a suitably calibrated instrument, *Nimbus 7*, was launched in 1978. Its success led to the hurried installation of a similar instrument on board the Solar Maximum Mission, late in the planning stages of that satellite. Ironically, the results from that instrument have been more widely reported than the results from the instruments that study solar flares.

Figure 14.4 shows the results of 153 days of solar monitoring reported by R. C. Willson. During six solar rotations, several large sunspot groups are carried past our view. Clearly, large sunspot groups may decrease the solar luminosity by about 0.1 percent.

No effect on climate is expected when a single sunspot group passes by, because the temporary decrease in luminosity is compensated probably within a few weeks, and the Earth's weather probably does not respond to such a small change that quickly.

Does the solar luminosity remain constant when averaged over times much longer than a month? Not so! From 1980 to 1986, the solar luminosity measured by the Solar Maximum Mission fluctuated owing to individual spot groups but decreased on average by 0.02 percent per year. The trend appears to have stopped with the solar minimum of 1986 and reversed. The Sun is becoming brighter as the activity of the new cycle increases.

If we may extrapolate from 5 to 80 years, the lack of sunspots during the Maunder minimum implies a slightly lower than-average solar luminosity during about 80 years. If such a decrease is real, is it long enough for the Earth's climate to respond with a little ice age? We do not know.

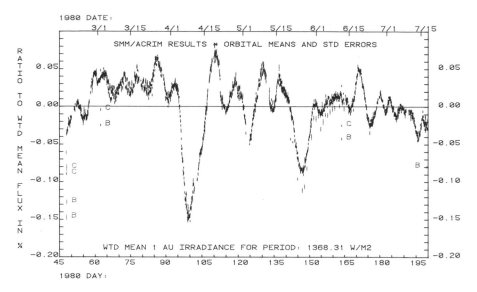

Figure 14.4. Spots and sunlight. Solar irradiance was measured on the Solar Maximum Mission during 153 days in 1980. Two 0.1 percent dips in irradiance are caused by large spot groups crossing the solar disk. The "zero" line is merely an average based on the expectation that energy temporarily held back by sunspots must emerge at some other time. (National Aeronautics and Space Administration, courtesy R. C. Willson and *Science*, vol. 211, 13 February 1981, pp. 700–702, copyright 1981, American Association for the Advancement of Science)

Is the Sun Shrinking?

Long-term changes in solar luminosity cannot be measured directly, but if they are caused by changes in the solar diameter, those changes might be inferred from historical data.

The first indications of solar shrinkage were obtained from visual measurements of solar size during the past century. Each measure is quite inaccurate, partly because of random errors of measurement, partly because of personal judgments on how to measure the solar size. But an attempt to remove these effects yielded a rather startling result: The solar diameter shrinks by about 1 part in 1,000 per century. If it continues unabated, the Sun should reach zero size in about 100,000 years!

The conservative and approved scientific response to such a

claim is an attempt to disprove it. The Sun's diameter can also be measured by the size of the Moon's shadow on the Earth during a total solar eclipse and by the time needed for the planet Mercury to cross the disk of the Sun. These measurements are more accurate, but they can be made only intermittently. The results imply that the solar diameter shrinks by less than 1 part in 6,000 per century. That is less than the previous result but still not very satisfying because, at the maximum rate allowed by this upper limit, the Sun would still reach zero size in less than a million years.

Perhaps the Sun "breathes" slightly, and any shrinking during the current millennium is only temporary. However, even an imperceptibly small "breath" might be important. A change in diameter by 1 part in 1,000 probably causes a change of luminosity by 1 part in 1,000, an amount that may well have long-term effects on the climate. Probably the Sun breathes much less than that, but the Solar Maximum Mission should teach us not to be too certain. Indeed, the ever more refined ground-based observations of solar rotation suggest some changes in solar structure, perhaps its convection, with time, and a corresponding change in size seems likely.

Is the Sun changing? Yes. Can the changes affect our climate? We should be prepared in case the answer is yes.

Chapter 15

The Solar-Stellar Connection: Is the Sun a Common Star?

The Nearest Star

The Sun dominates our daytime sky because it is so close to us. If it were at the distance of the next star, alpha Centauri, the Sun would appear to us as one of the brightest nighttime stars, slightly fainter than the star Sirius. If it were at the distance of typical bright stars in our sky, such as Betelgeuse, the Sun would be too faint to be seen through small telescopes.

The Sun is so near to us that we can recognize structural details on it. Good photographs can just resolve granules and other features with sizes merely a thousandth of the solar diameter. Modern technology is expected to do 10 times better. Nearly all other stars appear to us as small dots without structure, not only to the human eye but in all telescopes. One of the few exceptions is Betelgeuse, about a thousand times larger than the Sun. The largest telescopes, for instance the Multiple-Mirror Telescope in Arizona, using state-of-the-art electronics and computer analysis, can just barely resolve Betelgeuse. Its chromosphere is very extended, almost twice the size of its photosphere, so that it is very unlike the Sun.

The disparity in the appearance of the Sun and other stars has raised two questions.

Is the Sun a Star?

Historically, the significant question was, Is the Sun a star? Several philosophical arguments suggested the answer was yes, even during the Renaissance. This idea became more plausible when Galileo demonstrated the profusion of stars to be seen through a telescope. It became widely accepted when Newton argued that motions on Earth are the same as those beyond the Earth, in the solar system and presumably also among the stars. Nevertheless, proof that the Sun is a star remained elusive until the nineteenth century, when the first distance to a star could be measured and its luminosity and size could be computed.

Detailed proof appeared near the end of the nineteenth century, when stars were classified according to their spectra. The Sun fits neatly into one of these categories. It is classified as a G2V star, where the G2 designates the color and surface temperature of the Sun and the Roman numeral V implies that the Sun is in the hydrogen-burning stage of its life, the main and longest-lasting stage of stellar life (see chapters 1 and 2).

Most stars range in surface temperature from half to four times the Sun's temperature; in mass, from $\frac{1}{20}$ to 20 times the Sun's mass. Among hydrogen-burning stars, the luminosity ranges from a thousand times fainter to 10,000 times more powerful than the Sun. Lifetimes in the hydrogen-burning state vary from many billions of years, for stars with low luminosity, to a mere million years, for the most luminous stars. There are stars much larger than the Sun, such as Betelgeuse, but these are in a late and relatively short stage of their lives, taking a last forced gasp before expiration. There are also vastly smaller and less luminous stars, but these are dead stars in the sense that they have exhausted their nuclear energy supplies.

The Sun is an average star in that its properties fall close to the middle of the ranges occupied by the stars. It is a common star in that we know thousands of G2V stars. However, if a space traveler arrived at our part of the Milky Way Galaxy and started counting stars, he would be overwhelmed by smaller, cooler stars and he would find the Sun in the upper 10 percent of all the stars, no matter what property he listed them by.

Are Stars like the Sun?

Today, the important question is, Are other stars like the Sun? Do they have spots, chromospheric activity, cycles of activity, X-ray emission, coronae, flares, radio bursts, and so on?

Yes, there are stars with spots. Yes, there are stars with activity cycles. Yes, there are stars emitting X rays. So far, however, we can recognize most such stellar activity only if it is much more vigorous than on the Sun. Solar activity would be nearly undetectable if the Sun were at stellar distances, with the exception of its X rays. The *Einstein* Observatory (see chapter 8) could detect solar X rays, at least during periods of high activity, even if the Sun were at stellar distances.

Hot versus Cool Stars

Stellar activity comes in two forms, one in "cool" stars and one in "hot" stars.

The surface temperature of cool stars ranges from about 3,000 to 8,000 degrees. The Sun, with a surface temperature of nearly 6,000 degrees, is well within the regime of cool stars. Most cool stars probably feature activity like that of the Sun, even though we can observe only a very small fraction of it. Activity is also expected theoretically, because the cool stars have convection near the surface, and this convection presumably helps to create the magnetic field that then causes the stellar activity.

The surface temperature of hot stars ranges from about 8,000 to 30,000 degrees. Their surfaces have negligible convection, they rotate much more rapidly than most cool stars, and their activity is dominated by strong stellar winds. Whereas the present solar wind would take 50 trillion years to use up the mass of the Sun, the winds of some hot stars can remove a significant fraction of a star in as short as a million years.

Starspots

The radiometer on board the Solar Maximum Mission detected the weakening of sunlight due to sunspots by merely 0.1 percent. Such sunspots are still undetectable on other stars.

Fortunately, there are starspots much larger than solar spots. Large spots rotating into and out of view measurably change the brightness of some stars. Even the mere rotation toward the limb

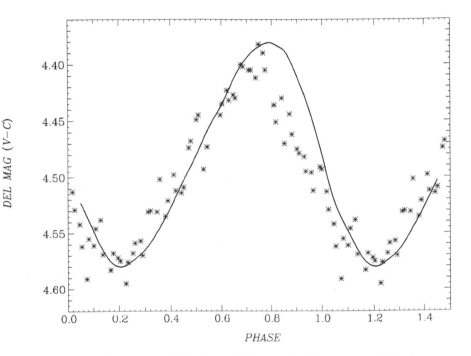

Figure 15.1. Starspots. With time, this star in the group called the Pleiades changes in brightness in a regular fashion, repeating every six hours. The crosses are the brightness data obtained during nine hours on a single night in 1983. The vertical scale may be read as percentage changes in brightness. The brightness variation by about 20 percent has been interpreted as being due to two spots, occupying about 8 percent of the stellar surface, and rotating with the star once every six hours. The solid curve summarizes the data obtained two years earlier. The difference between 1981 and 1983 suggests a slow migration of the spots. The light curve has changed again since the time of these observations. The data were taken with the 0.7-meter telescope of the Flower and Cook Observatory. (Courtesy J. Stauffer and *Astronomical Journal*)

is noticeable, because the spots then appear foreshortened and small. Figure 15.1 shows an extreme example. The data are consistent with two spots occupying about 8 percent of the stellar surface. The star is a very rapid rotator, a property related to its youth. Rotation periods of several days are much more common.

The spots in the star of figure 15.1 appeared to migrate with time, but they remained present for at least several years, for otherwise the average brightness would have changed with time. If sunspots lasted for as many stellar rotations, they would last well over a century and be reminiscent of the red spot on Jupiter. But even long-lived starspots fade and are replaced by others. The result is a gradual variation in the average brightness of a star, as in figure 15.2. Optimists may recognize a 60-year cycle of activity.

Stellar astronomers assume that starspots are dark and a long-term decrease in luminosity means an increase in spot size. They have not yet reckoned with the Sun's becoming slightly fainter at sunspot minimum, in an amount similar to the darkening by a single spot group (see chapter 14).

Chromospheric Activity

Measuring brightness is not an efficient way to detect activity, neither on the Sun nor on the stars. Fortunately, the Sun shows us more effective ways. Imagine monitoring the entire Sun at the wavelength of 6563 Å, the wavelength of hydrogen used to make the spectroheliograms of figures 4.4 and 11.1. The solar brightness at this wavelength may double or triple in a few days when a large active region appears. This relatively large variation in brightness of chromospheric radiation is a major advantage over the small changes in the luminosity of the star. However, this kind of observation uses only a small amount of the starlight. Therefore, one requires both bright stars and large telescopes. Currently, several hundred stars are being monitored to determine their activity.

Stellar Rotation

An immediate benefit of watching stellar activity is that stellar rotation can be measured directly. Rotation is of great interest because the magnetic dynamo probably depends on rotation, and activity depends on the magnetic dynamo.

Are the more rapidly rotating cool stars more active? Indeed,

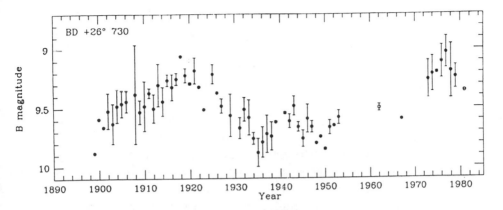

Figure 15.2. Changes in stellar spots. This very cool star rotates with a period of only 1.9 days and has slowly changed in brightness over the years. At its faintest, when it is presumably most spot covered, the star is only about half as bright as it is at its brightest. Activity seems to occur in a cycle of about 60 years. The bars on the annual data points indicate both real changes in star brightness during the year and errors of measurement. The older data were obtained from the collection of photographic plates at Harvard University. (Courtesy Sallie L. Baliunas; reproduced, with permission, from the *Annual Review of Astronomy and Astrophysics*, vol. 23, copyright 1985 by Annual Reviews Inc.)

among cool stars with nearly the same surface temperature, the more rapid rotators have stronger chromospheric emission, that is, stronger activity. However, the relation between rotation and activity is different for different surface temperatures. Some additional factor is indicated.

Dynamo Again

Our notions of the dynamo (see chapter 7) suggest what this additional factor might be. The dynamo depends not only on rotation but also on convection. The vigor of convection in the outer regions of the star is closely related to its surface temperature. Convection is most efficient for stars somewhat cooler than the Sun.

A simple measure of convection efficiency is the time needed for an eddy to turn over, that is, the time needed for hot gas to rise from the bottom to the top of an eddy and for cooled gas to sink. Work on the terrestrial atmosphere showed a ratio to be very useful, namely the ratio of the Earth's rotation period to the atmospheric eddy turnover time. This ratio has been named the Rossby number.

Robert Noyes of Harvard University investigated how the stellar activity depends on the Rossby number, defined now as the ratio of stellar rotation period to eddy turnover time. The eddy turnover time depends on surface temperature. Figure 15.3 is his very reasonable result: Fast rotation measured according to the Rossby number implies high activity, no matter what the surface temperature.

Figure 15.3 was the expected result based on ideas about the dynamo. Does it confirm these ideas? Only in a very general sense: It confirms that rotation and convection are important ingredients. However, no theory comes even close to predicting figure 15.3.

Differential Rotation?

The solar dynamo depends not only on rotation but also on differential rotation, the fact that the solar equatorial regions rotate once in a shorter period than regions at higher latitudes. Are stars also rotating differentially? If the two spots on the star of figure 15.1 reside at different latitudes, can we observe their different rotation periods?

If we monitored the Sun from afar and if active regions existed at the equator and at 60 degrees north, then we would have to monitor the Sun for well over a year in order to recognize the slight differences in rotation rates between the two active regions. No individual active regions last that long. Instead, new active regions might arise, say, 90 degrees in longitude ahead of the faded ones at 60 degrees north. If we monitored the Sun, incorrectly assuming a permanent active region at 60 degrees north, we would deduce a slightly wrong rotation rate for that latitude, and we would deduce a substantially wrong value for

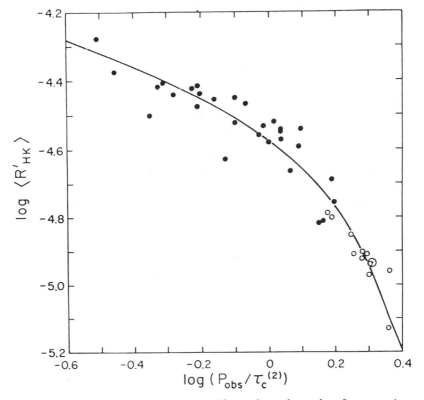

Figure 15.3. Activity and rotation. These data show that fast rotation implies high chromospheric activity. The vertical axis is a quantity that on the Sun would measure the area of active regions. The horizontal axis indicates the Rossby number, the ratio of rotation period to eddy turnover time in a star. Both quantitites range over a factor of about eight. (Courtesy R. W. Noyes and *Astrophysical Journal*)

the difference in rotation rates at the equator and at 60 degrees north. Evidently, careful and long-term stellar observations are required to demonstrate differential rotation.

Four Sun-like stars are deemed to show differential rotation large enough and consistent enough that it cannot be "explained away." These stars are among the more rapidly rotating and the more active stars. Once again, the observations confirm the ingredients for the dynamo and stellar activity: convection, rotation, and differential rotation.

Stellar Cycles

Several dozen stars have now been monitored for their chromospheric emission over a span of about 10 years, and a few have been sampled over almost 20 years. Are such intervals long enough to recognize stellar activity cycles?

Figure 15.4 shows the results for four stars. It is easy to believe that HD 81809, a G2V star like the Sun, has a regular cycle of 8.4 years. The analysis is much more difficult for HD 190007, which has rather large irregular activity superposed on some apparently more regular cycle. In this case one can make good cases for periods of 3.4, 5, and 9 years, or some combination thereof.

If we had monitored the solar chromosphere for a 20-year interval at various times during the last three centuries, we would have found cycle periods anywhere from 5 to 15 years. Some of this range occurs because the solar cycle is not strictly 11 years long but varies between about 8 and 13 years. Some of the range occurs because the cycle variation is rather irregular (see figures 6.1 and 6.2) and the data over any 20-year interval can be misinterpreted. Therefore, we should be careful not to exaggerate the stellar results based on 10 or even 20 years. If a star is said to have a cycle of 8 years, for instance, we may rejoice in the detection of a stellar cycle of roughly solar duration. But we should make no claims as to its regularity or persistence over the next few decades.

If there are stars with much longer cycles, we would not yet have detected them.

Where to Monitor the Stars

To obtain data on stellar rotation one must make nightly observations for many nights in a row, without interruption by clouds. To obtain data on stellar activity cycles one must monitor stars over many years. For both kinds of observations, telescopes must be dedicated to these tasks. The telescopes at the National Observatories are not available for such purposes. Privately operated observatories are required. Olin C. Wilson and Arthur H. Vaughan used the 60-inch telescope on Mount Wilson

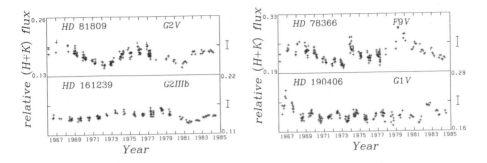

Figure 15.4. Stellar activity cycles? Changes in chromospheric emission over an 18-year interval were recorded for four stars with a surface temperature near the solar value. The two stars on the left are relatively inactive compared to the ones on the right. The star HD 81809, a G2V star like the Sun, has a fairly regular activity cycle. The best value for the cycle period is 8.4 years. For the slightly warmer HD 190406, the best value is about 2.6 years. The eye can discern cyclical variations for the other stars, but one should not try to find "a stellar cycle." (Courtesy Sallie L. Baliunas; reproduced, with permission, from the *Annual Review of Astronomy and Astrophysics*, vol. 23, copyright 1985 by Annual Reviews Inc.)

for many years, while Mount Wilson was operated by the Carnegie Institution of Washington. The Carnegie Institution has now withdrawn its support, having decided that a new telescope in the southern hemisphere has higher priority for its financial resources. Interim support is provided by the National Science Foundation. It remains to be seen whether long-term support can be arranged for the 60-inch telescope. The annual costs are not very large when compared to other scientific ventures, but during a time of very limited scientific budgets the competition is still intense. A possible alternative to Mount Wilson may be Lowell Observatory in Flagstaff, Arizona, which has a more modern spectrograph but only a 40-inch telescope.

Stellar Activity

Given that starspots and stellar active regions have been well established, one may well expect to discover other examples of stellar activity just like that on the Sun.

The observations of X rays from cool stars by the *Einstein* Observatory (see chapter 8) suggest that most cool stars have a corona rather like that of the Sun, with the X-ray emission concentrated over active regions. Just as starspots may be larger than sunspots, stellar coronal X-ray emission may be stronger than that of the Sun. Some cool stars may be nearly covered by an X-ray-emitting corona.

Flare stars may brighten dramatically. But stellar flares can be detected photographically only if they are orders of magnitude more intense than even the largest solar "white-light" flares. The flares can be detected more easily in the X rays, in ultraviolet spectral lines, and by their radio emission. Special campaigns have been mounted to observe a few stars that are the most likely to flare. One campaign was quite successful. Figure 15.5 shows the results. Clearly, the flare radiates in many ways. The time scales are reminiscent of solar flares. In general, the most energetic radiations have the shortest duration.

Stellar flares are most easily detected by their radio emission. The detectors have been developed the most, and observing time is available on large radio telescopes. The recent observation of flare radio "spikes" on the Sun, highly polarized and lasting only 0.1 second, has led observers to try to detect similar spikes from stars. Kenneth Lang, of Tufts University, used the large collecting area of the 300-meter radio telescope in Arecibo, Puerto Rico, to observe stars with a time resolution of 0.2 seconds. Figure 15.6 shows one radio flare with spikes. The high polarization of the spikes indicates emission by fast electrons in strong magnetic fields associated with the flare.

Coronal X rays, flares, radio spikes, and many more features all indicate that other stars are indeed similar to the Sun. But we observe only the most intense activity. The many additional phenomena seen in detail on the Sun should remind us not to oversimplify our notions of other stars.

The Young Sun

The Sun is a middle-aged star, about 5 billion years old. Was it different when it was young? The question is vital for the evolu-

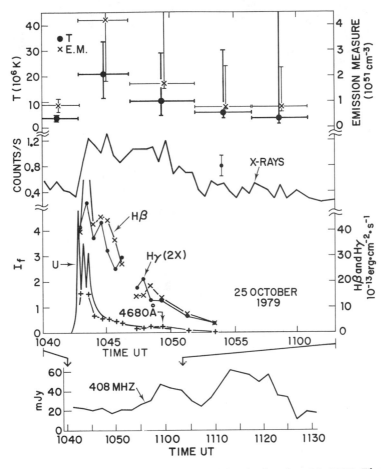

Figure 15.5. Stellar flare on YZ Canis Minoris, October 25, 1979. These coordinated observations in many wavelength bands show some of the complexity in attempting to interpret stellar flares. The horizontal scale of the main panel covers 20 minutes. The temporal behavior of the flare depends on the observing instrument, just as for solar flares. The curve marked X rays shows 30-second averages of the data from *Einstein* Observatory. At the top are two interpretations of the data, namely, the temperature and a measure of the amount of radiating gas. The highest temperature (dots) is about 20 million degrees, similar to that in solar flares observed with only 30-second resolution. The scales marked H-beta and H-gamma show the brightness of the flare in the light of hydrogen, observed at McGraw-Hill Observatory; those marked U and 4680 Å show stellar brightness observed at Cloudcroft Observatory. The radio emission at 408 megahertz came from Jodrell Bank, United Kingdom; only the radio peak at 1100 Universal Time is stellar, the still later peak is unexplained. (Courtesy S. Kahler and *Astrophysical Journal*)

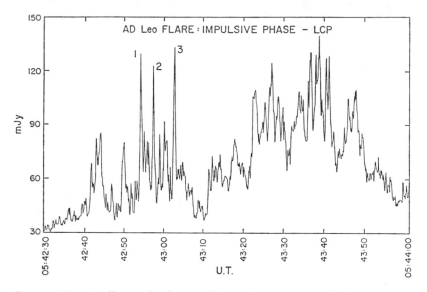

Figure 15.6. Stellar radio burst. The radio intensity of this star was measured with very high time resolution at the 300-meter radio telescope in Puerto Rico. The data shown span a 1.5-minute interval during a radio burst lasting about 25 minutes. The three numbered "spikes" are similar to spikes observed during solar flares. The sharpest spikes on both the Sun and this star last less than 0.2 seconds; so brief a period was unexpected merely a dozen years ago. (Courtesy Kenneth R. Lang and *Astrophysical Journal*)

tion of life on Earth, because life is critically dependent on liquid water, and liquid water, in turn, is possible on Earth only if the solar illumination is appropriate.

The earliest life detected on Earth occurred about 3.4 billion years ago, merely 1.2 billion years after the Earth formed. Therefore, liquid water certainly existed on Earth when both the Earth and the Sun were only about a quarter of their present age. Presumably liquid water existed even earlier while these simple life forms evolved. The presence of liquid water that long ago is somewhat surprising. Models of solar evolution make the young Sun appear only very slightly cooler but about a third less luminous than now. If the young Earth then received a third less radiative heat, simple estimates suggest that water should have been frozen back then, contrary to the detected evolution of life.

Perhaps the puzzle is solved by recognizing that the Earth's atmosphere was quite different back then. At the very least, oxygen did not exist in the air then. This notion is attractive but so far difficult to substantiate.

Another solution may depend on the Sun. What if solar activity back then was much more important, so important that it influenced the structure and temperature of the Earth's atmosphere? No evidence for such activity is preserved in terrestrial rocks. However, if the Sun is a common star, then such activity should be observable on young stars. Indeed, young stars show enormous activity.

Flare Stars

Most flare stars are very young stars. Some 469 cool flaring stars clustered in the Pleiades all are about 50 millions years old. Many flaring stars, named after their prototype T Tauri, are caught in the evolutionary stage when central nuclear fusion has just turned on, when they "settle down" to the main and long-lasting hydrogen-burning stage of their lives.

If young stars flare frequently, they should also show considerable chromospheric activity. Indeed, they do. Furthermore, if they are very active, they should rotate rapidly, according to figure 15.3. Indeed, most of the fast rotators among the cool stars are young. The prototype for the young stars, T Tauri, rotates once every 2.8 days.

How long does youthful activity last? Enough young stars of known ages have been observed to measure how the average spin rates decrease with increasing stellar age. Stars less than about 300 million years old have slowed very little. About half the rotation may be gone after 1 billion years. Judging from the Sun, most of it is gone after 5 billion years.

The Young Active Sun?

If young stars are very active, the Sun may once have been more active than now. How much more active was it during its first billion years? Judging from the stellar averages, it may well have rotated five times faster during its first billion years. Its chromospheric activity may have been five times higher. The rate of strong flares depends sensitively on chromospheric activity. Five times higher activity may imply a huge increase in the rate of strong flares. The young Sun may have frequently had enormous flares, far exceeding the energy of even the largest flares observed today, similar to the outbursts seen in flare stars today!

Lunar and Meteoritic Confirmation

Flare activity during solar youth is confirmed by dust and rocks from the Moon, materials whose ages are known as parts of lunar history. Energetic particles from solar flares buried themselves in the lunar dust and rock and left tracks that can still be observed in an electron microscope. One finds that lunar rocks exposed to the Sun some 3 billion years ago were exposed to energetic particles from flares much like the energetic particles produced by flares today. However, these data cannot be used to find the frequency of particle-producing flares at that time.

Still earlier, large flares must indeed have been dramatically more powerful than now. Meteorites contain grains with nuclear reaction products caused by energetic particles from solar flares. The grains must have been hit by the particles before the grains were incorporated in the meteorite. The accumulation of the grains into meteorites is thought to have occurred when the Sun and Earth formed, during an interval of only a few hundred million years. The data suggest that energetic particles were a thousand times more numerous then than they are today. Thus the meteorites record for us signs of powerful solar flares during the very earliest portion of the Earth's evolution.

Surely the early evolution of life on Earth was strongly influenced by the activity on the young Sun. Observing young stars may teach us something about the young Sun's activity. However, that activity would have affected the Earth's atmosphere in ways that we cannot yet even estimate. Life on the early Earth may remain a puzzle for a while longer.

Chapter 16

The Solar Chimes

Star Tests Again

Our understanding of the Sun and solar evolution could be improved substantially if we could develop a new observational method to probe the solar interior. In particular, solar activity depends on convection as deep down as one-third of the solar radius. That convection is not directly observable and theoretically is still poorly understood. Any new information about the convection zone would be extremely valuable.

Solar structure and evolution depend on the abundance of helium relative to hydrogen, a quantity that is difficult to measure in the solar atmosphere. A precise knowledge of the helium abundance is significant for many other astronomical topics, from the death of stars to the origin of our Universe. Any new probe of the solar interior would provide an immediate new measure of the abundance of solar helium.

One method of probing the interiors of some stars, though not of the Sun, has been known for over half a century.

Stellar Pulsations

The brightness of some stars is observed to change in a regular fashion. The stars are deduced to grow and shrink periodically, to pulsate. The best known of the pulsating stars are named after their prototypes, RR Lyrae and Cepheids. Their periods of pulsation range from about half a day to several weeks. The pulsations are not merely atmospheric phenomena. Some stars may change their diameters by one-third. Surely the entire star must participate in such an upheaval. The pulsation provides us with a means of probing the entire star.

The radial pulsations are confirmed by stellar spectra. Careful measurement of the wavelengths of the absorption lines show that the wavelengths change slightly with time, typically by roughly 1 Å. The changes are quite regular and predictable and their period is the same as that of the changes in brightness.

The change in wavelength is interpreted as a motion of the stellar surface, specifically of the stellar surface visible to us, facing us. A slightly decreased wavelength is interpreted as the stellar surface approaching us; a slightly increased wavelength is interpreted as the stellar surface receding from us. The change in wavelength due to motions is known as Doppler shift.

When the timing of changes in wavelength and brightness is compared, the visible stellar surface is seen to approach us just when the changes in brightness tell us that the star is growing. This is consistent with radial stellar pulsation.

The most precisely measured property of a pulsating star is its period of pulsation. Many stars have now been observed for decades. Stars observed for hundreds of pulsations have their period measured to substantially better than 1 percent, a degree of accuracy that is rare in astronomy. An accurate period of a pulsating star, together with frequent simultaneous measurements of changes in brightness, color, and motions, provide the information needed for a detailed modeling of the pulsation throughout the star. The observations and the models will agree only if all parts of the interior, including the convection zone and the helium abundance, are modeled correctly. Thus the pulsations provide a test for our understanding of the stellar interiors, of their helium abundance, and of our ability to model convection.

Analogy: Sounds from an Organ Pipe

The period of pulsation of a star can be compared to the notes produced on an organ pipe. Why, for example, does an organ pipe produce middle C, which is a vibration in air with a period of $\frac{1}{400}$ of a second? The organ pipe contains air. When air is stirred in any part of the pipe, the resulting changes in pressure travel as sound waves, causing motions of air everywhere else in the pipe. Most of this noise simply dies out. An audible note is produced only if all these motions can somehow be organized, for instance, is a sound wave travels the length of the pipe during exactly one period of the wave. Such a wave provides the "fundamental" tone of the pipe. If a desired pitch has been chosen for the fundamental, such as middle C, the length of the pipe is chosen accordingly. The lower the fundamental pitch, the longer the pipe.

Audible notes are also produced for sound waves that travel down the pipe during exactly two periods of the waves, or three periods, and so on. These waves provide the "harmonics" of the pipe, at higher pitch than the fundamental. Middle C might also arise as the harmonic of a longer pipe with a fundamental of lower pitch. Thus, if we hear middle C we learn a lot about the organ pipe but not everything. It helps to hear an entire organ concert.

The analogy between pulsating stars and organ pipes is a fairly sound one. Stars also pulsate in an orderly fashion, and the pulsation periods are such that a sound wave can cross the star in one pulsation period (fundamental), or in two periods (first harmonic), or three, and so on. Indeed, most pulsating stars are observed to pulsate in the fundamental or in the first harmonic, and some stars pulsate in both.

Stars differ from organ pipes in that they do not have a uniform sound speed. The sound speed depends mainly on the temperature. It is fastest near the stellar center, relatively slow far out. The run of sound speed with distance from the center of the star determines just how long it takes a sound wave to cross the star. When we have observed the pulsation period precisely, we can deduce the average of the sound speed throughout the star. That value will agree with the computed value if we have assumed the correct mass and chemical composition of the star and used the correct model for convection. Therefore, these properties are also tested.

On the whole, stellar models and pulsation observations agree remarkably well. Nevertheless, for some pulsating stars the comparison has forced scientists to revise the adopted mass of the stars or the abundance of helium relative to hydrogen.

All the pulsating stars are aging stars that are larger and hotter than the Sun. The cause for their pulsation is reasonably well known. According to both observation and theory, the Sun should not pulsate. Indeed, the Sun clearly does not change its radius by even 0.1 percent on any time scales between minutes and months. Such pulsations would be very easy to detect. Nevertheless, the Sun does pulsate. It pulsates in thousands of overtones. The Sun provides a veritable stellar concert, which in principle can be used for thousands of tests of solar structure.

Five-minute Oscillations

Some 20 years ago, an unexpected phenomenon was discovered on the Sun. R. B. Leighton, R. W. Noyes, and G. W. Simon decided to investigate the solar surface motions by taking frequent spectroheliograms adjusted to detect up-and-down motions in the photosphere. To their surprise, spectroheliograms taken about five minutes apart looked alike. Many granules seemed to be jointly bobbing up and down in a regular fashion, with a period of about five minutes, in addition to their individual relatively random motions. This discovery started an entirely new line of solar investigations into the "solar five-minute oscillations."

The five-minute oscillations can now be detected in many other ways. One way is to measure the up-and-down motions of individual granules. These motions produce tiny changes in the wavelengths of absorption lines. Figure 16.1 shows the spectrum of a quiet part of the Sun. The "wiggly lines" are absorption lines distorted by the wavelength shifts of many granules. Since granules last only a few minutes, the pattern of wiggles at first sight appears random. But figure 16.2 shows an underlying regularity. The granules seem to bob up and down together for a while, then the order disappears, later it reappears, and so on. Typically, the bobbing motions have a period of about five minutes, but there is substantial variation. The ordered motions are quite

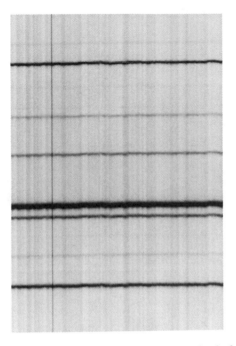

Figure 16.1. Spectrum of granules. A slit admits the light from a long, thin section of the Sun (figures 5.3 and 5.4), including many granules. The light is then dispersed into its colors (see plate 6) and a small range in wavelengths (vertical) is recorded here, including four prominent (horizontal) absorption lines and several weaker ones. The right side of the picture samples light from one end of the slit, the left from the other end. Each vertical line is the light from one granule. The absorption lines "wiggle" up and down because some gas in the granules is rising, shifting the absorption to shorter wavelengths (up in the figure); other gas is sinking, shifting the absorption to longer wavelengths. As expected, the brighter, hotter gases are rising, bringing new energy to the solar surface, and the cooled-off darker gases are sinking. The granule velocities are about 0.5 kilometer per second. (National Optical Astronomy Observatories)

slow, no more than about 200 meters per second, at most half the granular motions.

When Leighton and his colleagues discovered the five-minute oscillations, they could measure velocities as low as 10 meters per second. Astronomers observing stars consider this undreamed-of accuracy. Yet it was not enough to investigate the five-minute

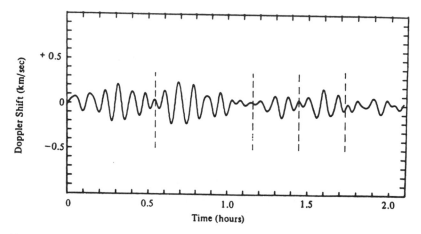

Figure 16.2. Five-minute oscillations. Wavelength shifts and velocities, recorded for about two hours at a single point of the Sun, show intervals lasting about an hour, during which the velocities oscillate regularly with a period of close to five minutes. (Courtesy Oran R. White.)

oscillations further. A significant technical improvement became available in the early 1970s. At that time, a "Fourier tachometer" was built that incorporated a direct wavelength calibration in the laboratory. This made it possible to measure changes in velocities on the Sun as small as 0.5 meter per second. For comparison, a common walking speed on Earth is about 1.5 meters per second (5.4 kilometers or 3.4 miles per hour).

The Tones of a Bell

It is now clear that the Sun does pulsate, but motions in the Sun are three-dimensional, not merely radial as in the pulsating stars, and the oscillations occur at thousands of frequencies. A better analogy than the one-dimensional organ pipe is a bell: It is two-dimensional and sounds many tones simultaneously. Among the possible tones on a bell, some tones are very pure and sharply tuned. They shake the bell in a geometrical pattern that fits the shape of the bell. Other tones are of low quality. The latter can be minimized if the bell clapper hits the bell at just the right places.

The best solar analogy is a combination of organ pipe and bell. Solar pulsations are really sound waves that must "fit" in three dimensions: They must fit with their radial motion as the waves do in an organ pipe and in pulsating stars. The waves must also fit onto a great circle around the Sun, as on a bell.

Plate 15 shows one possible pattern of motion that might be observed on the Sun if we could instantly detect it throughout the Sun. If we watched for a few minutes, all the motions would regularly reverse their directions so that we could also measure a period of oscillation.

Organ pipes and bells can produce thousands of notes. We have learned to excite only a modest number of tones when we expect to hear music. The Sun is not very musical in this sense. Thousands of patterns like those in plate 15 occur simultaneously, each with its own tone. For still unknown reasons, most of the excited tones have periods ranging from about two to eight minutes. Each tone causes velocities at the solar surface of merely about 0.1 meter per second, or less. The thousands of simultaneous oscillations may add up for a while, and then one observes the five-minute oscillations of figure 16.2. Sometimes they cancel, which explains why the five-minute oscillations of figure 16.2 disappeared occasionally.

Frequencies and Patience

If we are to learn something about the Sun from its concert, we must resolve that concert into its separate tones. Figure 16.3 shows the kind of resolution in frequency that is of interest; higher resolution is much better.

The solar tones are very pure and highly tuned. Some frequencies are precise to 1 part in 10,000 and differ from frequencies of other tones by only 1 part in 10,000. To measure frequency to an accuracy of 1 part in 10,000 one must observe the Sun continuously for at least 20,000 oscillation periods. At five minutes per oscillation, this amounts to two months.

How can one observe the Sun continually for many days without the intervention of nighttime or bad weather? In 1981, two groups managed to solve this problem. One group went to the

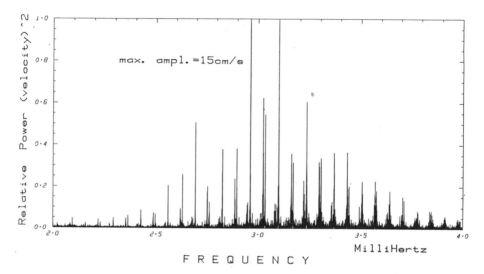

Figure 16.3. Solar oscillation frequencies were measured over a period of three months from Hawaii and from Tenerife, Spain. The Sun oscillates at hundreds of highly tuned frequencies between 2.5 and 3.8 millihertz, corresponding to periods between 6.7 and 4.5 minutes, respectively. The strongest mode has a velocity amplitude of merely 0.15 meter per second (0.5 kilometer per hour). Radial velocities were averaged over the entire solar disk. Within each group of modes, the frequency separation between modes informs us of the deep solar interior. (Courtesy G. Isaak, University of Birmingham)

Antarctic, where the Sun could be observed continuously except in periods of bad weather. They succeeded in completing five days of uninterrupted observations. Another group combined observations from Hawaii and Tenerife, Spain. The two sites are far enough apart in longitude that the Sun was observable for an average of 22 hours per day. Figure 16.3 was made in this fashion. Another pair of stations is operated at the Big Bear Solar Observatory in California (see figure 11.2) and at an observatory in Israel.

A network of solar observing stations will soon be established. The project is called the Global Oscillations Network Group (GONG). Currently, the quality of observations is being tested at 11 sites with generally good weather located all around the Earth. Some combinations of six sites observe the Sun for better than 97 percent of the time. It all goes well, six sites will receive a specialized instrument that will measure the solar oscillations in a nearly automated fashion. The data will all be transferred to the National Solar Observatory. The combined data should be equivalent to continuous, uninterrupted measurements for two years at a single station. The total cost of the project is about 15 million dollars, including site testing, construction, operation, and data analysis, which is to be completed by 1996. The European space probe *Solar and Heliospheric Observatory* may extend these data during the late 1990s. It will detect solar oscillations that cannot be recognized by the GONG method of observations.

Once a basic understanding of the solar oscillations is achieved, one may again use the observations of a single station and correct the observations for the inevitable interruptions by nighttime, weather, and equipment maintenance. Thus a relatively simple monitoring of solar oscillations may be carried out for several solar cycles. Already there is a hint that some oscillation frequencies change with the solar cycle.

Angular Patterns

In view of thousands of observed and millions of observable modes, exact frequency information is not enough to identify any one mode, just as middle C will not identify an organ pipe. The angular pattern should also be observed. In practice, every experiment averages patterns over some part of the solar surface. It averages most patterns to zero and, therefore, is able to detect only restricted sets of modes. For instance, some experiments measure the velocity averaged over the entire solar disk. Such experiments can distinguish only modes with rather few wavelengths spanning a great circle. Modes with short wavelengths are simply averaged out. Figure 16.3 was obtained in this way.

The GONG experiment, with its high frequency resolution, will recognize many more modes than is possible with the experiment of figure 16.3. Nevertheless, its insensitivity to short wavelengths is a significant limitation. Other experiments average over the north-south direction and detect only modes with patterns resembling the shape of a peeled orange. Figure 16.4 was obtained this way. That figure was expressly constructed to emphasize the short-wavelength modes. Experiments with still less angular averaging are now operating at individual observatories. Although they do not yet span an uninterrupted time base of sufficient length, they have already shown that short-wavelength modes are different near sunspots. With time, such observations may inform us about the structure of sunspots far below the photosphere. This subject has been dubbed "solar tomography."

Results on Helium, Convection, Rotation

Some of the most accurate data on solar oscillations appear in figure 16.4. In principle, if we could construct an accurate model of the Sun, we should be able to match all the observed oscillations. On the whole, the agreement of observations and predictions is very good, but the interesting information is contained in the small discrepancies. These take time to sort out. So far, some major controversies have been settled concerning the outer third of the Sun. Results for deeper portions are still debatable.

One major astrophysical controversy has to do with the amount of helium in the Universe. The abundance of helium relative to hydrogen not only influences stellar structure and evolution, but it also provides a significant clue to the first few minutes of the Big Bang origin of our Universe. Helium has been difficult to measure spectroscopically, but a ratio of 1:3 of helium to hydrogen by mass has been generally agreed upon. Data on solar oscillations agree with this ratio. Already they clearly exclude a ratio of 1:4 and narrower limits of error are expected to be established soon.

One significant question, never quite answered by stellar models, is, What is the depth of the convection zone? Small changes in a stellar model, for instance in the assumed efficiency with

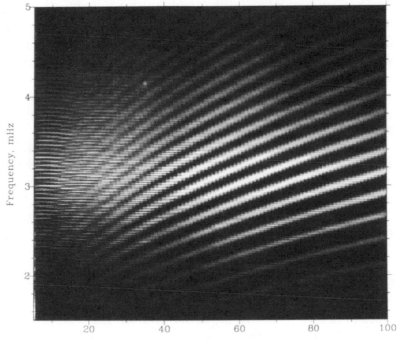

Figure 16.4. Solar oscillations (white) derived from measurements in California and in Israel. Detected wave frequencies (vertical) correspond to wave periods between about 4 and 8 minutes. The horizontal axis is the number of waves that fit around the solar equator. For each number—say, 50—waves are observed at several frequencies, corresponding to various overtones of an organ pipe. The data of figures 16.3 and 16.4 complement each other in informing us about convection and rotation in the solar interior. (Courtesy K. Libbrecht, California Institute of Technology; and *Astrophysical Journal*)

which convection carries heat, may imply large changes in the theoretical depth of the convection zone. The observed solar oscillations confirm the depth expected from the most favored models, about one-third of a solar radius. Thus the observations help to confirm the assumptions that are needed to describe the convection.

Does the Sun rotate more rapidly inside? Measurements of solar oscillations can provide information on the interior solar rotation. Waves of any one mode travel more easily with rotation than against it. Consequently, waves of one mode traveling in opposite directions acquire a slight difference in frequency. The frequency difference has been detected and analyzed, at least for some modes. The period of rotation is nearly constant with depth through the convection zone. Its behavior further toward the center is under debate. There is no evidence for any drastic changes in rotation period near the solar center.

Ultimately, one would like to measure not only rotation but also differential rotation in the interior. Such information would provide much insight into the generation of magnetic fields and solar activity. This requires observing surface patterns more like those in plate 15, patterns that have been averaged out in figures 16.3 and 16.4. Preliminary results suggest no dramatic change in differential rotation through the convection zone.

To fully analyze solar oscillations will be a complicated task. In part it involves some basic mathematical difficulties that may prevent researchers from obtaining precise answers. Another difficulty is that at the same time we must improve our knowledge of many aspects of the Sun, convection being only the worst of many problems. Suitable methods of analysis are now being prepared by a team of experts from nine countries.

The Solar Clapper

Bells are made to ring by a clapper. Why does the Sun ring when everyone expected it to be static? The reason is probably to be found in solar convection. Convective motions are quite irregular. Frequently, gas masses move toward each other and collide. The collision causes the emission of sound waves, noise. Most of the noise just dies away. Some sound waves move outward to heat the chromosphere (see chapter 8). But some of the sound waves "fit" the Sun so that they can travel many times around the Sun, picking up strength along the way. They are amplified to become the observable solar oscillations.

The tonal quality of a bell depends not only on the shape and material of the bell but also on how the clapper strikes the bell. Analogously, careful observation of the solar tonal quality will provide information on the dynamics of convection inside the Sun. One has to imagine that the oscillations are constantly made and then destroyed again. Each oscillation lives a few days or weeks. The longer it lives, the sharper is its frequency. One can already tell that some solar notes are very sharp, some not. That is the analogy to the tonal quality of a bell. Much effort is now going into measuring it.

Stellar Oscillations

The solar oscillations averaged over the entire disk can in principle be detected on other stars. If their amplitudes are similar to those on the Sun, one needs a large telescope with a high-quality spectrograph available every night for a long enough period to distinguish the frequencies of the most important modes. Suitably large telescopes arc now too much in demand for this purpose. Perhaps a specially dedicated telescope is needed. Of course, the oscillations on other stars may be an order of magnitude stronger, requiring much less observational effort. Indeed, about a dozen stars show regular fluctuations in brightness with amplitudes up to about 1 percent. At least one of them shows 12-minute oscillations equivalent to the solar 5-minute oscillations, with regular spacing in frequency equivalent to that of figure 16.3. However, these are not common stars. They are warmer than the Sun and strongly magnetized. Their oscillations may be caused by magnetic phenomena rather than by convection. Therefore, they may not tell us anything about solar-type oscillations on more ordinary stars.

The reward of successfully observing stellar oscillations will be great. We will be able to compute how solar oscillation frequencies have changed and will change as the Sun evolves. Observation of analogous stellar oscillations will provide the first direct measurement of stellar ages, known so far only by deduction. A direct measure of stellar ages will help to improve our understanding of the formation of planets and of the theory of stellar evolution. It will be crucial for determining the ages of the oldest stars. The oldest stars, in turn, figure in cosmology since these stars had better be younger than the age of the Universe!

There is some indication that this subject may not be an easy one to tackle. Starting about a decade ago, S. Vogt observed the details of stellar absorption lines among hot stars rotating much more rapidly than the Sun. To his surprise, the observations of some stars could be explained by a corrugation traveling across the stellar surface. The corrugation resembles a single very strong solar oscillation. The large amplitude may be a good omen for the observation of many more stellar oscillations. But the main surprise is the appearance of only a single mode. The selection of just one mode may be due to the fast rotation. In any case, observation of only a single mode limits the information we might gain about the stellar interior.

Stars can oscillate in many other ways, depending on their gravity and magnetization. Even the Sun appears to oscillate in a mode with a period of 160 minutes. Lively debate surrounds both the reality of the mode and its possible interpretation. Is the 160-minute variation an artifact of the Earth-based observations? Most likely, but if it is truly solar and caused by the buoyancy of the interior gases, why is there only one mode?

Quantitative Measurements

The measurement of solar oscillations is one of the few accurate astronomical measurements we can expect to make. Preliminary results have already confirmed previously uncertain values of the helium abundance and of the depth of the solar convection zone.

Within a few years solar observations may provide a basic understanding of highly turbulent convection. Such an understanding will test and extend our knowledge of terrestrial convection and will shed light on the solar dynamo and the origin of solar activity.

Chapter 17

Before You Can Analyze Your Results

Solar Science Became Big Science

Solar astronomy has benefited from the work of many individuals. George Ellery Hale, known for developing major astronomical instruments, pioneered the discovery of magnetic fields in sunspots. Robert C. McMath and the privately sponsored McMath-Hulbert Observatory in Michigan spearheaded the work in solar cinematography and, following the advances in military science during World War II, in the mapping of the infrared solar spectrum.

The term "big science" is generally applied to the research efforts after World War II, when scientific projects grew beyond the scope of an individual scientist working in his laboratory. In that sense, some solar science has always been big science. Eclipse expeditions have always been major undertakings. A century ago, an eclipse expedition consisted of a voyage of several months and required at least a ship and crew. However, an energetic individual could still mount the expedition, make the observations, and publish the results afterwards.

Today, an eclipse expedition is likely to be sponsored nationally and be conducted by teams of scientists, each of whom is expected to concentrate on relatively specialized instruments and to collaborate in evaluating their data.

Solar science now incorporates much space research, which is necessarily part of big science. Satellites like the Solar Maximum Mission (figure 17.1) require tremendous financial support not only because a launch is expensive, but also because every one of the numerous instruments on such a mission has to be

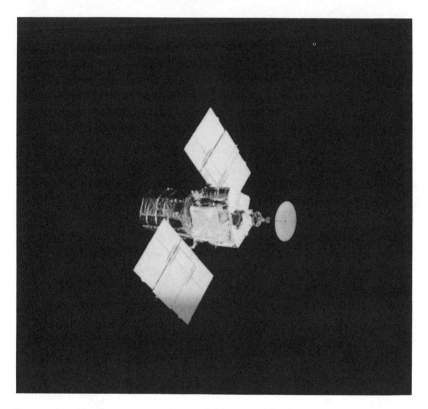

Figure 17.1. Solar Maximum Mission (National Aeronautics and Space Administration)

designed, built, and tested by a team of specialists. Moreover, ever since launch in 1980, the Solar Maximum Mission has required teams of experts both to maintain operations and to determine daily the scientifically most useful observations.

Interdisciplinary Science

Solar science has also become highly interdisciplinary in scope. It is very difficult for an individual scientist to isolate an important and solvable problem merely by personal ingenuity. Expertise from other branches of science is essential. Even the lone theoretician solving some mathematical problem must maintain a broad perspective of solar science, and especially of the observations, if the work is to be considered valuable.

An enormous variety of physics enters into solar questions. Consider for example, what observations would be useful to distinguish whether X rays or electron beams cause the sudden evaporation of the chromosphere beneath a flare (see chapter 11). To answer this question we require expertise ranging from hydrodynamics, plasma physics, nuclear and atomic physics to electronics, optics, and X-ray detection. Cooperation among scientists with many backgrounds is vital. Not surprisingly, solar science has attracted many people trained in physics. The traditional emphasis on the description of solar phenomena has been replaced by an emphasis on "solar physics."

Owing to the interdisciplinary nature of solar science, workshops have become an important forum for discussing ideas and exchanging information. Many are national, some international. Typically the workshops last three days. They end with a "homework assignment" for each participant, to be accomplished before the next workshop. No homework, no participation. No attendance at workshops, no participation.

Who Becomes a Solar Scientist?

Workshops can be readily accommodated because most solar research occurs at large research-oriented institutions such as the National Solar Observatory (at Tucson, Arizona, and Sacramento Peak, New Mexico), the High Altitude Observatory (a part of the National Center for Atmospheric Research at Boulder, Colorado), the NASA Goddard Space Flight Center (Greenbelt, Maryland), the Naval Research Laboratory (Washington, D.C.), and the Harvard-Smithsonian Center for Astrophysics (Cambridge, Massachusetts). These centers have facilities for instrument development and for frequent visitors and are able to stay at the forefront of the science.

The emphasis on workshops has a major practical consequence. It encourages solar scientists to devote their full time to solar research. Rather few solar scientists are to be found at universities, and these tend to be supported by federal grants so that they have few or no teaching duties.

The lack of emphasis on teaching means that few graduate students are being trained in solar physics. Only a handful of doctoral graduates have entered the field in recent years, and they have been trained at only five major universities. Consequently, many junior research positions are difficult to fill. The average age of solar scientists is increasing even more rapidly than that of physical scientists in general.

The U.S. solar community is well aware that too few solar scientists are being trained. How can more scientists be attracted to this field? One remedy is to invite (through member contributions) promising students to attend solar meetings. Another is to focus on quantitative observations that will attract space physicists, plasma physicists, and even mathematicians who are accustomed to the precision required in experimental science. Indeed, the methods needed to analyze solar oscillations have attracted geophysicists and even an oceanographer.

Nevertheless, there is a practical problem, one that solar science has in common with all of the space sciences: U.S. students perceive career risks in this discipline because the work depends heavily on space experiments and thus on year-to-year political decisions. There is certainly some justification for that view. The more stable European community of solar scientists is quite able to attract numerous excellent students. The NASA administration is quite aware of the need for stability. But the larger a space project the more difficult it is to provide the needed stability.

NASA continually supports modest projects such as data analysis, the development of theory to explain the data, or the development of new instrumentation. Certainly NASA has great interest in the use of the enormous amounts of data collected from the many satellites and space probes that have already been launched. Data from Skylab and the Solar Maximum Mission are freely available. Individual university scientists may apply for grants in the range of $10,000 to $80,000, mainly to cover the salary needed to analyze the data. Also supported are the costs of computing and publication, the cost of attending workshops, and "indirect" charges by the university such as those for maintaining a research library. Proposals to NASA for such projects do not demand a huge effort. A proposal is usually quite specific. The scientist can write it in a few days at most. Peer review and NASA personnel's selection process may occupy a few more days of work. The chance that a proposal will be accepted is typically about a third. Support from NASA is sometimes complemented by support from the National Science Foundation for ground-based observations. This kind of project is usually very suitable for training graduate students.

Spaceflight Experiments

The planning of a spaceflight experiment is probably the largest task for many solar scientists. Such an experiment might revolve around one of the X-ray spectrometers on the Solar Maximum Mission. Shepherding such an experiment from conception to final results may take up a large part of a scientists's career. It involves far more than merely making decisions about what is to be measured.

An important step in planning a space experiment is to convince people that it is worth the investment. To the productive solar scientist, this aspect of science is as important as the results, for without one you cannot achieve the other. In fact, the promotional aspect may take substantially more time than the final analysis of the results.

The proposal for a flight experiment is submitted following the "Announcement of Opportunity" by NASA, which indicates that a flight such as the Solar Maximum Mission is under consideration. In fact, the proposal will have been written weeks earlier because the mission has long been anticipated and the instrument to be proposed has been discussed for months with the aid of a separate small grant. Otherwise, there would not be sufficient time to plan the instrument between the time of the NASA announcement and the deadline for the proposal.

Soon after the deadline, panels of scientists review the proposals to determine how effective the various instruments are expected to be. What are their capabilities? What important questions can they address? Can the instruments be built reliably within a reasonable budget? Later other panels will evaluate the balance of the instruments in terms of the goals of the mission. They may select a preliminary list of instruments for further study. Financial support is then provided for detailed planning and initial construction. As few as 1 proposal in 10 may reach this stage. Even some of these instruments may later be combined with others or dropped altogether. Meanwhile the entire mission must run the gauntlet of approval all the way up to Congress.

The prototype and later the actual flight instrument must survive extensive testing to ensure success in space after the rigors of launch and deployment. Testing must be particularly rigorous for launch in the Space Shuttle. Each experiment must also be integrated into the mission's scheme of data transmission and steps must be taken to prevent the instrument from causing electrical interference in any of the other instruments on the mission.

If all goes as planned, a flight experiment will involve some $10 to $30 million and may occupy an entire group of scientists and technicians for several years. Rather few universities are capable of supporting such a group, and most of them started participating early in the space age and grew with it.

New Start of a Major Mission

Major missions are even more complicated. First the idea must be investigated to determine whether there is sufficient initial interest in it. Then it must be endorsed or approved by a variety of advisory panels and committees. Does the mission have a clear and significant goal? Does it enhance the overall goals of space exploration? Financial support is then provided for an extensive study of the science, instrumentation, and cost. Then the competition starts in earnest. Might the mission compete scientifically and/or financially against some other mission within the same division of NASA? More generally, how does it compare with top-priority missions identified in oceanography, in the life sciences, in planetary exploration, or in high-energy astrophysics?

Clearly, the choices cannot be made purely on scientific grounds. The goals of NASA and its perception of what the public will support necessarily enter the evaluation. The first such decisions are frequently made at the NASA Goddard Space Flight Center, later ones at NASA Headquarters, always with ample advice from individual scientists and advisory committees.

Once NASA decides on its proposals for "new start" missions, the documents are sent to the Office of Management and Budget. That office may decide the mission does not fit into the present administration's priorities or it may request a substantial "de-scoping" for resubmission with a lower budget. If a mission still survives, it moves on to Congress, where it undergoes budgetary review by the authorization and appropriations sub-committees and committees of both the House and the Senate. Finally, it must be approved by the full Congress and by the president.

Nearly all missions of the early 1980s were deemed to require launch by the Space Shuttle. Therefore, the missions had to respond to the political realities of the shuttle program, in particular to the delays in the shuttle launches, with the first launch postponed to April 1981 and subsequent launches at intervals much longer than anticipated. An early plan for an extensive series of shuttle-launched "Spacelabs" was reduced drastically by the cancellation of 37 planned experiments. Two Spacelabs were successful, one of which had a strong solar component. The third was lost with the *Challenger* explosion.

Two solar space missions, the Solar Maximum Mission (SMM) and the International Solar Polar Mission (ISPM), illustrate extremes in the outcome of the actions needed to support solar science.

Solar Maximum Mission (SMM)

The cost of the Solar Maximum Mission came to about $200 million for planning and construction plus about $20 million per year for operating it. The SMM must be counted as an enormous success. Its Announcement of Opportunity was published in 1975. The mission was launched by a Delta rocket on February 14, 1980. Five years is considered an almost incredibly brief time. The experiments were designed to observe details of major flares. The Sun obliged, providing a remarkably varied assortment of interesting major flares during the next few months.

After a simple part in SMM failed, the telescopes could no longer point accurately at desired parts of the Sun. Unlike most parts in most satellites, this one had no backup. Most of the SMM instruments became useless. Fortunately, the SMM happened to be the first satellite equipped for possible subsequent space repair. Repair by shuttle promised many "firsts" including first shuttle capture of a free-flying spacecraft, first operational use of the Manned Maneuvering Unit, first use of extravehicular power tools, first extensive checkout of a spacecraft berthed in the shuttle, and more. What is more, SMM became famous because of the add-on experiment measuring the solar energy flux and the effect of sunspots on this flux (chapter 14). The fame of SMM made possible many educational contacts between solar scientists and members of the congressional staffs. All this led to congressional approval of the SMM repair mission.

The SMM repair was accomplished during a flight of the space shuttle *Challenger* (April 6–13, 1984). That mission cost an additional $40 million, to which must be added the cost of the shuttle launch. NASA estimates that a shuttle launch costs roughly $100 million, but the estimate depends on how one assigns the costs of the NASA launch and communications facilities and other infrastructure.

Several of the major experiments on SMM are still working and still observing the Sun. NASA still spends about $7 million a year on SMM, about half of which goes toward supporting university scientists and about half to operations. In fact, the SMM may be the only working U.S. solar mission in space during the next solar maximum in 1990. Perhaps it can be refurbished in the 1990s. However, a very strong sunspot maximum would raise the Earth's atmosphere enough so that SMM would reenter the atmosphere and be destroyed as early as 1989.

International Solar Polar Mission (ISPM)

Decades of patience are the hallmark of the organizers of the International Solar Polar Mission (ISPM), currently renamed Ulysses. Scientifically, the goals were excellent. The Earth, being nearly in the plane of the solar equator, samples only a tiny fraction of the solar wind. From Earth, one obtains only highly foreshortened and uncertain views of the solar polar regions. Yet we know that the polar regions are distinctive. The first signs of a new solar cycle appear near the poles. High-speed solar winds may reach the Earth, with harmful consequences, even if they derive from coronal holes rather near the poles.

ISPM was to consist of two space probes. One was to travel to Jupiter, use its gravity for a boost out of the ecliptic plane, travel toward the Sun, fly over its northern polar regions, swing southward around the Sun, then fly over its southern polar regions before again departing for the outer solar system. The other probe was to have a similar orbit but first was to travel south and then swing north. At the same time, similar solar-observing equipment was to be placed in orbit about the Earth. Thus the three sets of cameras, X-ray detectors, magnetometers, and other equipment that would be in operation were expected to yield a three-dimensional view of the solar wind, of the extended corona, and of the coronal holes, and at least a two-dimensional view of the polar regions. Optimists conceived of a second orbital swing past the Sun that would collect additional data about seven years after the first, but the final design did not include sufficient power to point the instruments toward the Sun for this opportunity.

One of the two space probes was to be planned and built by the European Space Agency, the other one by NASA. Both were to be launched by NASA. Each group was to hold an open competition for experiments, with U.S. proposals eligible for the European probe and vice versa. The progress of ISPM is outlined in Table 3. Plans were well under way when the U.S. craft was canceled. Revised plans were again well under way when the *Challenger* disaster occurred. On the current (March 1988) schedule, results from ISPM can be discussed no earlier than 23 years after initial development!

International Collaboration

The cancellation of the U.S. ISPM craft not only reduced the scientific benefits but had considerable political consequences.

Table 3. International Solar Polar Mission—Ulysses

Year	Stages of development

1960 First serious interest in out-of-ecliptic mission.
1973 Orbits via Jupiter technically feasible; develop and test advanced technology and techniques.
1975 Symposium at NASA Goddard Space Flight Center; advisory committees and study group write mission document.
1977 Announcement of Opportunity.
1979 "New start" approved by Congress; instrument simulation and design, fabrication, test and calibration.
1980 Cancellation of U.S. craft as savings measure; Europeans have already invested heavily in their craft; redesign of experiments to make best use of single craft; U.S. experiments on European craft survive.
1983 Integrate instruments into spacecraft; spacecraft tests.
1986 *Challenger* disaster preempts May launch.
1987 Compete with Jovian probe "Galileo" in scheduling for the only shuttle large enough for either probe and for the single annual launch opportunity toward Jupiter; decision to launch at the earliest in October 1990.

If schedule is maintained
1991 December, swing past Jupiter.
1994 July, first arrival over a polar region.
1994 December, closest approach to Sun, at a distance still somewhat larger than Sun–Earth distance.
1995 June, arrival over other polar region.
1996 Earliest discussion of results, 23 years after first development.

The Europeans base their work on rather dependable five-year plans. They have difficulty comprehending the U.S. system, which requires even long-term programs to run the financial gauntlet of Congress every year. The ISPM cancellation raised many questions about the wisdom of collaborating with NASA, especially since the Europeans were planning their own launch facility, Ariane.

The stage for renewed scientific collaboration was set during a conference in 1982 in which groups of U.S. and European investigators exchanged outlines of how their respective decision-making and financial systems work. In principle, international collaboration has now come back into favor with the U.S. administration. Not only does science benefit from the exchange of information but projects can be undertaken at a lower cost to the countries involved. Currently, the budgets for solar-terrestrial space research and for the space station are strongly influenced by the opportunities to collaborate with European and Japanese counterparts of NASA. In practice, rules against exporting U.S. technology slow down negotiations and have already prevented collaborations with counterparts in the Soviet Union.

The Japanese have already provided important solar data through their satellite *Hinotori*. They will fly the satellite *Solar A* during the next solar maximum with detectors for hard X rays and gamma rays. A team of U.S. scientists will participate and provide a telescope to observe soft X rays. Japanese spacecraft have flown reliably and exactly according to their long-range plan. If the U.S. scientists are slowed down by unexpected reductions in funding, they may well be left behind at the launch pad.

The European Solar and Heliospheric Observatory may be the next solar spacecraft with a wide range of experiments. NASA is participating actively by supporting much of the instrument development, especially for measuring solar oscillations, and is planning to launch the spacecraft. NASA will also receive the data by continuous coverage with the Deep Space Network (which is a major concession), and will oversee the experiments from the Goddard Space Flight Center.

High Spatial Resolution

The success of the Solar Maximum Mission led the solar community to push for a telescope in space with a spatial resolution down to a scale of about 100 kilometers. This resolution is deemed necessary to properly measure the magnetic flux tubes and other small magnetic structures that generate flares and most other solar activity. For a while the planned mission was called the Solar Optical Telescope, designed to fly on the Space Shuttle. It was de-scoped in 1985. Its subsequent reincarnation, called High Resolution Solar Observatory, was shelved in the summer of 1987. A new design has been requested, suitable for a

free-flying satellite launched by a Delta rocket. It is to be called
Orbiting Solar Laboratory and is to cost an estimated $500 mil-
lion. Many solar scientists would have preferred this all along,
because Space Shuttle flights last only about a week and regularly pass through the Earth's shadow, which is not very condu-
cive to observing the development of an active region or a flare.

Meanwhile, the European solar community started planning
the ground-based Large European Solar Telescope. It is designed for accurate measurements of the polarization of the solar light because, as almost always, polarization informs us of
magnetic fields (see chapters 5, 12). With the added participation of the United States and China, it has been renamed the
Large Earth-Based Solar Telescope. Its 2.4-meter mirror can
theoretically resolve 100 kilometers on the Sun. However, one
must first contend with the distortions caused by daytime atmospheric turbulence. Two apparently very quiet sites are ocean-
surrounded mountains on the Canary Islands and on Hawaii.
Nevertheless, the desired high spatial resolution will probably
demand continual electronic control of the mirror shape in order to compensate for atmospheric distortions. The entire project is likely to cost about $40 million. The United States may
bear up to 30 percent of the cost.

The ground-based telescope will have a resolution similar to
that planned for the Orbiting Solar Laboratory; it will be
cheaper, and it will be built more quickly. Will the former obvi-
ate the need for the latter? Probably not. The ground-based telescope will have its optimum spatial resolution over only a very
tiny two-dimensional field of view. The space-based telescope
will have a larger field of view and will be sensitive to ultraviolet and x-radiations, and thus to a third dimension, height in the
solar atmosphere. It will investigate the same phenomena but in
an independent fashion, so that the two telescopes should complement each other admirably.

The complementary nature of space- and ground-based observations is exemplified by an experiment on *Spacelab 2*, aboard
the space shuttle *Challenger*, from July 29 to August 6, 1985. Its
solar cameras recorded "exploding granules," apparently ordinary solar granules that seem to explode and send large, expanding brightness fronts through the surrounding granules. One sus-

pects that our description of the granulation is quite inadequate without taking the exploding granules into account. Since they are intermittent phenomena, their further observation requires the continuous observing opportunities only available on the ground. Now, a Swedish solar telescope on La Palma in the Canary Islands has been instrumented to take many electronically recorded images per minute and automatically select the best frames. Exploding granules are observed routinely, with a spatial resolution twofold better than that achieved on *Spacelab 2*.

The prime facility in the United States for improving high-resolution solar observations is the National Solar Observatory. It operates the McMath Solar Tower at the Kitt Peak National Observatory, Arizona (see figure 5.3), and several instruments on Sacramento Peak, New Mexico, including a major spectrograph at the Vacuum Solar Tower (see figure 17.2). The Vacuum Solar Tower is used to test flexible mirrors so that they can receive further development. Also, with computer-corrections, the photographs taken there have become so sharp that some magnetic flux tubes have been identified and followed briefly in their motions. This research is supported by the Air Force and by the Lockheed Missiles and Space Company.

The annual operating budget for the National Solar Observatory is about $3.5 million, which includes the expenses for developing new instrumentation. In addition, the budget includes about $2 million per year for the Global Oscillations Network Group, until that project is completed.

These ground-based solar budgets may seem small compared to the cost of space experiments, but this comparison is inappropriate because the observatory is not part of NASA. The National Solar Observatory is affiliated with, and receives about an eighth of the budget of the National Optical Astronomy Observatory, which is funded by the National Science Foundation. To maintain the solar fraction of the larger budget, solar physicists must compete vigorously with stellar astronomers, who greatly outnumber solar astronomers. Of course, both solar and stellar astronomers must also join forces to argue for a larger budget for all of them, one level higher up at the National Science Foundation.

An essential component of solar ground-based research in-

Figure 17.2. The Vacuum Solar Tower at Sacramento Peak, New Mexico. The evacuated light path down the tower eliminates the worst of the atmospheric distortion. A mirror consisting of 19 electronically controlled segments eliminates most of the rest. (National Optical Astronomy Observatories)

volves radio observations. The maps of coronal magnetic fields near sunspots and flares (see figure 12.5) cannot be obtained in any other way. The major solar radio facilities at the Very Large Array, New Mexico, and at the Owens Valley Radio Observatory, California, represent still more competitors for funds from the National Science Foundation. Improvements in receiver technology have been clearly identified to permit much more detailed measurements of magnetic fields near flares during the next solar maximum.

Visibility of Solar Science

Solar astronomers are few in number, and are grouped in relatively few institutions. Most of them feel that it is essential to make this science highly visible, lest funds for solar research dry up. Many of them see a very vigorous solar program pursued in Europe and in Japan and sense that U.S. solar science is losing its momentum.

A particularly frustrating example of the lack of visibility occurred on September 13, 1985. The U.S. Secretary of Defense was "absolutely delighted" on that day that an antisatellite test had successfully destroyed a "dead" Air Force satellite called P78-1. Although the main goals of this satellite had been accomplished, a coronagraph aboard P78-1 was still regularly taking white-light photos of the extended corona. Its images extended to 10 solar radii, compared to the 6 for the Solar Maximum Mission, so that it could trace coronal structure and coronal transients (see figure 9.3) much further out from the Sun. Moreover, the data were of great value because they extended over almost seven years, which is a substantial segment of a solar cycle. The photos had even shown four comets approaching the Sun and disappearing in the corona. The appropriate scientists had been given no opportunity to present the value of continuing observations before the satellite was destroyed.

Solar science is by no means alone in needing and seeking visibility. On the contrary, the health of science requires that the various sciences compete vigorously. Vigorous competition forces each science to make explicit its contributions to the other sciences, to the physical quality of life, and to the intellectual quality of life. Solar science can do well in this respect.

Solar science historically provided the first realization that the Sun has spots and that, therefore, celestial bodies may not be "perfect." A solar spectrum produced the first evidence of helium.

Solar science will soon contribute to quantitative tests of our understanding of neutrinos (see chapter 2), of convection that is vastly more turbulent than achieved in the laboratory (see chapter 16), and of electrical heating and associated gas dynamics on a scale vastly beyond the reach of laboratories (see chapters 8 and 12).

Practical benefits abound. Monitoring of the development of active regions (see chapter 4) and especially of changes suggesting impending flares (see chapter 12) is essential if we are to understand and alleviate the industrial effects of geomagnetic storms, the radiation damage suffered by sensitive satellites and by future astronauts, and the fate of satellites during solar maxima. Will the billion-dollar Hubble Space Telescope survive the next solar maximum?

Only solar science can help us to understand many of the long-term changes happening on Earth (see chapter 14), changes that grow increasingly important as the Earth's population spreads ever further into regions where living conditions are marginal at best. Space experiments have convinced us that the solar radiations, particles, and winds all play a role. Measurements are needed to quantify that role over several decades, starting now.

The Sun is the only star seen in any detail. It suggests to us the complexity of active regions on other stars (see chapter 15). It tells us that a corona and a stellar wind can jointly surround a star. It alone provides information without which we might totally misinterpret the spectra of other stars.

The Sun shows us that acceleration of particles to high speeds occurs very easily, even in rather modest flares (see chapter 13). Thus it provides confidence that the more cataclysmic events observed on other stars and galaxies can be interpreted as bigger and better versions of nearly daily phenomena.

Intellectual Quality of Life

The measurement of solar oscillations lays the basis for the first direct measure of the ages of other stars (see chapter 16), and thus of the time since the Big Bang, the origin of our Universe. The neutrino measurements (see chapter 2) will focus on our understanding of basic physics and of cosmology. Overall, the Sun exemplifies for us many of the processes that have shaped the evolution of the Universe, of the stars, of the elements, and even of life on Earth.

Finally, modern solar science portrays the Sun not as a perfect and unchanging ball of fire, but as an incredibly dynamic and ever-changing object. It is truly a restless Sun.

Subject Index

Pages in boldface indicate extended treatment.

Name Index